Information and Instructions

T0260120

This shop manual contains several sections each covering a specific group of wheel type tractors. The Tab Index on the preceding page can be used to locate the section pertaining to each group of tractors. Each section contains the necessary specifications and the brief but terse procedural data needed by a mechanic when repairing a tractor on which he has had no previous actual experience.

Within each section, the material is arranged in a systematic order beginning with an index which is followed immediately by a Table of Condensed Service Specifications. These specifications include dimensions, fits, clearances and timing instructions. Next in order of arrangement is the procedures paragraphs.

In the procedures paragraphs, the order of presentation starts with the front axle system and steering and proceeding toward the rear axle. The last paragraphs are devoted to the power take-off and power lift systems. Interspersed where needed are additional tabular specifications pertaining to wear limits, torquing, etc.

HOW TO USE THE INDEX

Suppose you want to know the procedure for R&R (remove and reinstall) of the engine camshaft. Your first step is to look in the index under the main heading of ENGINE until you find the entry "Camshaft." Now read to the right where under the column covering the tractor you are repairing, you will find a number which indicates the beginning paragraph pertaining to the camshaft. To locate this wanted paragraph in the manual, turn the pages until the running index appearing on the top outside corner of each page contains the number you are seeking. In this paragraph you will find the information concerning the removal of the camshaft.

CLYMER® MANUALS   **CLYMER**® PROSERIES    I&T SHOP MANUALS™

More information available at Clymer.com
Phone: 805-498-6703

Haynes Publishing Group
Sparkford Nr Yeovil
Somerset BA22 7JJ England

Haynes North America, Inc
859 Lawrence Drive
Newbury Park
California 91320 USA

ISBN-10: 0-87288-085-0
ISBN-13: 978-0-87288-085-6

© Haynes North America, Inc. 1983
With permission from J.H. Haynes & Co. Ltd.

Clymer is a registered trademark of Haynes North America, Inc.

Printed in Malaysia
Cover art by Sean Keenan

All rights reserved. No part of this book may be reproduced or transmitted in any form or by any means, electronic or mechanical, including photocopying, recording or by any information storage or retrieval system, without permission in writing from the copyright holder.

While every attempt is made to ensure that the information in this manual is correct, no liability can be accepted by the authors or publishers for loss, damage or injury caused by any errors in, or omissions from, the information given.

SHOP MANUAL
JOHN DEERE

SERIES 4030, 4230, 4430 & 4630

Tractor serial number located on rear of transmission case. Engine serial number located on front right side of engine block.

INDEX (By Starting Paragraph)

CONDENSED SERVICE DATA

GENERAL	4030 Gasoline	4030 Diesel	4230 Gasoline	4230 Diesel	4430 Diesel	4630 Diesel
Engine Make			OWN			
Engine Model	6303G	6329D	6362G	6404D	6404T	6404A
Number of Cylinders			6			
Bore-Inches	3.86	4.02	4.25	4.25	4.25	4.25
Stroke-Inches	4.33	4.33	4.25	4.75	4.75	4.75
Displacement-Cu. In.	303	329	362	404	404	404
Compression Ratio	7.6:1	16,7:1	7.5:1	16.8:1	15.5:1	15.5:1
Induction*	N-A	N-A	N-A	N-A	T	T-I
Cylinder Sleeves			WET			
TUNE-UP						
Firing Order			1-5-3-6-2-4			
Valve Tappet Gap–						
Exhaust-Inch	0.022	0.018	0.028	0.028	0.028	0.028
Inlet-Inch	0.014	0.014	0.015	0.018	00.018	0.018
Ignition or Injection Timing–						
Static	.....	TDC	.....	TDC	TDC	TDC
Running	24 at 2500	.....	20 at 2000	.....	.....	.....
Spark Plug Electrode						
Gap-Inch	0.025	.....	0.025	.....	.....	.....
Breaker Point Gap–						
Inch	0.020	.....	0.022	.....	.....	.....
Governed Speeds-Engine RPM–						
Low Idle	800	800	800	800	800	800
High Idle	2700	2660	2400	2400	2375	2375
Loaded	2500	2500	2200	2200	2200	2200
Horsepower at Pto						
Shaft**	80	80.33	100	100.32	125.88	150.66
Battery:						
Volts			12			
Ground Polarity			Negative			
SIZES-CAPACITIES-CLEARANCES						
Cooling system						
(Quarts)+	22	22	24	24	30	36
Crankcase Oil (Quarts)–						
Including Filters	17	17	12	17	17	17
High Clearance Final						
Drive Housing						
(Quarts)	2	2	2¼	2¼	2	.....
Transmission & Hydraulic System (Gallons)+ +						
Syncro-Range	13	13	13	13	13	23
Power Shift	.....	.....	11	11	11	13
Quad Range	13	13	13	13	13	23
Fuel Tank (Gallons)	35	35	37	37	56	65
Crankshaft Sizes and Clearances			See Paragraph 65			
Piston, Rings & Sleeves			See Paragraph 60			
TIGHTENING-TORQUES-Ft.-Lbs.						
Cylinder Head-Final	110	110	130	130	130	130
Main Bearing Caps			See Paragraph 65			
Connecting Rod Caps			See Paragraph 64			

*N-A = Naturally Aspirated; T = Turbocharged; T-I = Turbocharged and Intercooled

**Horsepower is according to Nebraska test except 4030 gasoline and 4230 gasoline models which are manufacturers estimate.

+ Add approximately 2 quarts if equipped with heater.

+ +Approximate capacity only. When draining, 3 to 6 gallons may remain in case. Add 4½-5 gallons if equipped with Power Front Wheel Drive.

FRONT SYSTEM

Three different tricycle front end units have been used: Single front wheel, Dual front wheels and "Roll-O-Matic" dual wheel tricycle units. Tricycle front end units require a special front support and steering motor assembly. Some units are convertible to adjustable axle models.

Adjustable front axle on tractors without front drive may be Low Profile, High Crop or standard height in narrow, regular or wide widths. Adjustable front axle on tractors with Power Front Wheel Drive is regular width only.

A Low Profile front axle with fixed tread width is available on some models.

TRICYCLE FRONT END UNITS

1. **REMOVE & REINSTALL.** The spindle extension (pedestal) attaches directly to steering motor spindle by four cap screws. To remove the unit, support front of tractor with a hoist using the special John Deere lifting plate (JDG-3) or other suitable support attached to tractor front support. Tighten the retaining cap screws to a torque of 300 ft.-lbs. when unit is reinstalled.

2. **SINGLE WHEEL TRICYCLE.** The fork mounted single wheel is supported on taper roller bearings as shown in Fig. 3. Bearings should be adjusted to provide a slight rotational drag by means of adjusting nut (4). The one-piece wheel & rim assembly (5) accommodates a 7.50"-16" tire, the two-piece rim (6) a 11.00"-12" tire. Wheel fork (1) is not interchangeable for the two tire sizes.

3. **DUAL WHEEL TRICYCLE.** An exploded view of the dual wheel tricycle pedestal and hub is shown in Fig. 4. Horizontal axles are not renewable. Service consists of renewing the complete pedestal assembly.

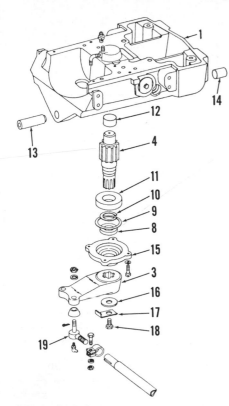

Fig. 1—Partially exploded view of front support that can only be used with adjustable and fixed tread front axles. Notice that axle pivot bosses are cast as part of support.

1. Front support	12. Bushing
3. Center steering arm	13. Pivot pin
4. Spindle	14. Bushing
8. Oil seal	15. Bearing quill
9. "O" seal	16. Washer
10. Snap ring	17. Lock plate
11. Ball bearing	18. Cap screw
	19. Rod end

Fig. 2—Partially exploded view of convertible front support assembly. The unit can be used with tricycle or adjustable tread axles. Axle pivot bracket (2) must be attached to support (1) to mount axle and steering arm (3) is bolted to the spindle (4). Tricycle fork (5) or pedestal (6 & 7) is bolted directly to spindle (4).

1. Front support
2. Axle pivot bracket
3. Steering arm
4. Spindle
5. Single wheel tricycle fork
6. Dual wheel tricycle pedestal
7. Roll-O-Matic pedestal
13. Pivot pin
14. Bushing
20. Ball bearing
21. Roller bearing
22. Spacer (some models)
23. Oil seal
24. Gasket
25. Steering motor spindle quill

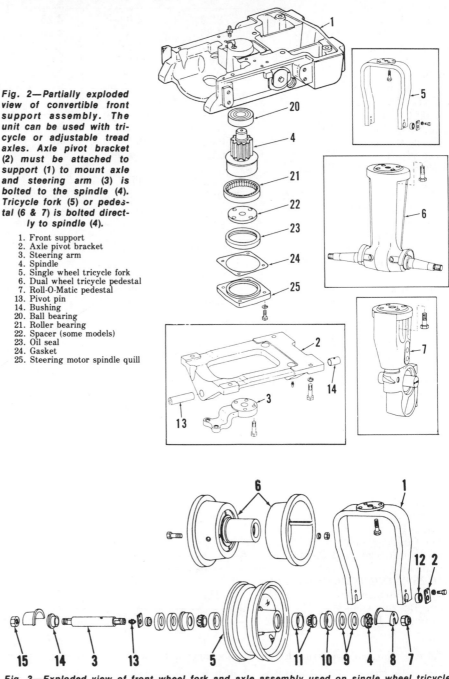

Fig. 3—Exploded view of front wheel fork and axle assembly used on single wheel tricycle models. The one-piece wheel (5) is for 7.50-16 inch tire, the two-piece rim (6) for 11.00-12 inch tire.

1. Fork	5. One-piece wheel	9. Felt washers	12. Washer
2. Lock plate	6. Two-piece wheel	10. Felt retainer	13. Grease fitting
3. Axle	7. Nut	11. Bearing cup and	14. Spacer
4. Bearing adjusting nut	8. Dust shield	cone	15. Nut

4. ROLL-O-MATIC UNIT. The "Roll-O-Matic" front wheel pedestal and associated parts are shown exploded in Fig. 5. The unit can be overhauled without removing the assembly from tractor.

Support front of tractor and remove wheel and hub units. Remove knuckle caps (6) and thrust washers (5), then pull knuckle and gear units (4) from housing.

The "Roll-O-Matic" unit is equipped with a lock (2) and lock support (3) which may be installed for rigidity when desired. Check the removed parts against the values which follow:

Knuckle Bushing ID . . . . . 2.127-2.129 in.
Knuckle Shaft OD . . . . . . 2.124-2.126 in.
Renew thrust washers (5) if worn.

Bushings are presized and contain a spiral oil groove which extends to one edge of bushing. When installing new bushings, use a piloted arbor and press bushings into knuckle arm so that OPEN end of spiral grooves are together as shown in Fig. 6. Bushing at spindle end of "Roll-O-Matic" unit should be pressed into arm so that outer edge (B) is 1/32-inch below edge of bore. There should be a gap (C) of 1/32 to 1/16-inch between bushings and distance (A) from edge of inner bushing to edge of bore should measure 3/16-inch. Soak felt washers in engine oil prior to installation. Install one of the knuckles so that wheel spindle extends behind vertical steering spindle. Pack the "Roll-O-Matic" unit with multipurpose type grease and install the other knuckle so that timing marks on gears are in register as shown at (M—Fig. 7). Tighten the thrust washer attaching screws to a torque of 55 ft.-lbs., and bend up corners of lock plates.

ADJUSTABLE AND FIXED TREAD AXLES

5. HOUSING & PIVOT BRACKET. The front axle attaches directly to the front support (1—Fig. 1) or to the pivot bracket (2—Fig. 2). On all models, install rear pivot pin bushing flush with the bottom of chamfer, and be sure that hole or lubrication channel is 180° from grease hole. Install front pivot pin flush with rear edge of pin hole in axle. Tighten pivot pin nuts to a torque of 220 ft.-lbs.

6. SPINDLES & BUSHINGS. Steering arm (6—Fig. 8, 9 and 10) is splined to spindle (5) and retained by a bolt. Spindle bushings are presized. Shim washers (8) are provided to adjust end play of spindle. Maximum allowable end play of spindle is 0.030 inch.

Refer to paragraph 35 for service to front axle used with power front wheel drive.

7. TIE RODS AND TOE-IN. The outer tie rod ends on axles with adjustable tread are adjustable with several holes provided. On all models, at least the inner ends are threaded. Make sure that tie rods are equal length so the

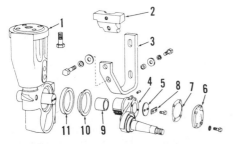

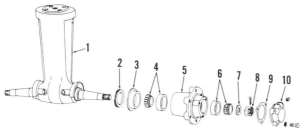

Fig. 4—Dual wheel tricycle pedestal showing one wheel hub and associated parts.

1. Pedestal
2. Oil seal
3. Oil seal cup
4. Bearing cup and cone
5. Hub
6. Bearing cone and cup
7. Washer
8. Nut
9. Gasket
10. Cap

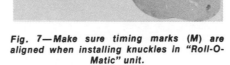

Fig. 5—"Roll-O-Matic" Spindle extension showing component parts. Lock (2) and support (3) may be installed to increase rigidity.

1. Pedestal extension	
2. Lock	7. Gasket
3. Support	8. Lock plate
4. Knuckle	9. Bushing
5. Thrust washer	10. Felt retainer
6. Cap	11. Felt washer

Fig. 7—Make sure timing marks (M) are aligned when installing knuckles in "Roll-O-Matic" unit.

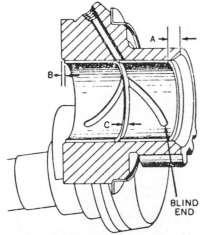

Fig. 6—Cross-sectional view of knuckle showing details of bushing installation. Refer also to paragraph 4.

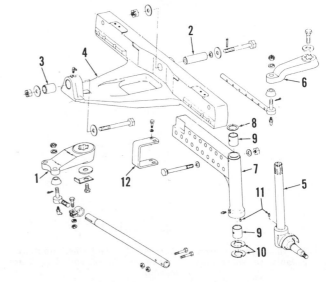

Fig. 8—Exploded view of typical, standard, adjustable tread front axle. The steering motor arm (1) is splined as shown in Fig. 1.

1. Steering motor arm
2. Pivot pin
3. Bushing
4. Axle housing
5. Spindle
6. Steering arm
7. Knee
8. Shim washer
9. Bushing
10. Thrust washers
11. Dowel pins
12. Clamp

tractor will turn the same in either direction. Adjust toe-in to 1/8 to 3/8-inch, and tighten clamp bolts to 35 ft.-lbs. torque. If tie rod ends are renewed, tighten nuts to 100 ft.-lbs. torque.

STEERING SYSTEM (EARLY TYPE)

Two different steering systems have been used. The early type covered in this section includes tractors before the following serial numbers:

4030 Models . . . . Before 008840
4230 Models . . . . Before 019559
4430 Models . . . . Before 027635
4630 Models . . . . Before 010647

All of these models are equipped with full power steering. No mechanical linkage exists between steering wheel and front unit; however, steering can be manually accomplished by hydraulic pressure when tractor hydraulic unit is inactive. Power for steering is normally supplied by the same hydraulic pump which supplies power for the brake system and hydraulic lift. A pressure control (priority) valve is located in rockshaft housing (or transmission cover if not equipped with rockshaft). Valve gives steering and brakes first priority on hydraulic flow.

Refer to paragraph 20 and following for service to later power steering system and components.

OPERATION

All Early Models Except 4630

11. The power steering system consists of the tractor hydraulic system described in paragraph 272, plus the steering control unit and motor described in this section. Refer to Fig. 15 for a schematic view of steering control unit and motor.

The control unit contains a double acting slave cylinder (2) on steering wheel shaft which is of approximately equal displacement to the cylinders in steering motor (9). Valve actuating

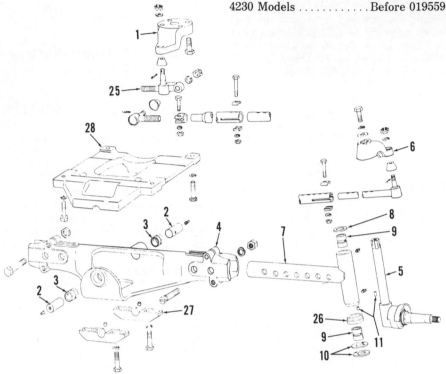

Fig. 9—Exploded view of Low-Profile adjustable front axle available on some models. Refer to Fig. 8 for legend except the following.

25. Inner tie rod connector
26. Dust shield
27. Pivot pin clamps with dowels
28. Axle pivot bracket

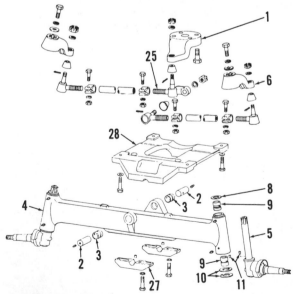

Fig. 10—Exploded view of Low-Profile fixed tread front axle available on some models. Refer to Fig. 8 and Fig. 9 for legend.

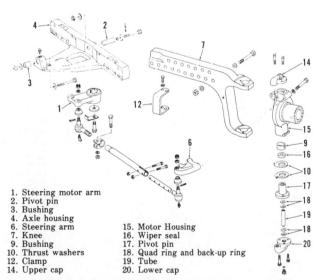

1. Steering motor arm
2. Pivot pin
3. Bushing
4. Axle housing
6. Steering arm
7. Knee
9. Bushing
10. Thrust washers
12. Clamp
14. Upper cap
15. Motor Housing
16. Wiper seal
17. Pivot pin
18. Quad ring and back-up ring
19. Tube
20. Lower cap

Fig. 11—Exploded view of adjustable tread front axle used on models with power front wheel drive. Refer to paragraph 30 and following for service to the drive motors and related parts.

levers (5) are in contact with operating collar (4) also mounted on steering shaft.

When the control unit is in neutral position, there is no fluid flow but oil at pump pressure is available at pressure line (6). When steering wheel is turned for a right or left turn, the steering shaft screw (3) will meet the resistance of slave cylinder (2) and shaft will move up or down on splines from the steering wheel shaft and will move collar (4) and levers (5) to open one pressure and one return needle valve. Fluid at pump pressure will then enter the control valve, lines and cylinders.

On a right hand turn, pressurized fluid is allowed to enter the bottom part of slave cylinder (2), fluid from top of cylinder is forced to left side of

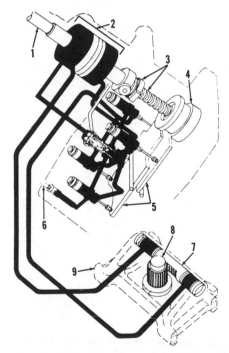

steering motor (9) and fluid from right side of steering motor (9) returns to sump. On a left hand turn, fluid at pump pressure is allowed to enter right cylinder of steering motor (9) and fluid from left cylinder is forced to top of slave cylinder (2). Fluid from bottom part of slave cylinder returns to sump through the open return valve. When turning in either direction, the slave cylinder meters the extent and speed of fluid flow and turning action, and is directly controlled by the steering wheel.

When no pressure is present at inlet pressure line (6), check valves prevent entry and discharge of fluid from steering lines. The piston in slave cylinder is manually moved by the steering shaft screw (3) and manual steering is accomplished by exchange of trapped fluid between slave cylinder (2) and steering motor (9).

Early 4630 Model

12. The steering motor (9—Fig. 15) on early 4630 model tractors differs from other models by using two operating pistons (7) to turn spindle (8). The added piston (7) is identical to the other one, and is used on the opposite side of spindle (8) to assist in turning either direction.

When making a right hand turn, pressurized fluid is allowed to enter the bottom part of slave cylinder (2), and at the same time fluid enters the right end of the rear cylinder to push rear piston to the left. As slave cylinder (2) is forced upward, the trapped oil at top of cylinder is directed to left end of front cylinder and forces front piston to the right to assist the rear piston in turning spindle (8) to the right. The trapped fluid at the left end of rear piston and the right end of front piston returns to sump.

On a left hand turn, pressurized fluid is allowed to enter the right end of front piston and left end of rear piston, which turns spindle (8) to the left. Trapped oil from the left end of front piston is directed to the top of slave cylinder (2) and forces piston down.

Trapped oil from the right end of rear piston and from below slave piston returns to sump. When turning in either direction, the slave cylinder meters the extent and speed of fluid flow and turning action, and is directly controlled by the steering wheel.

When no pressure is present at inlet pressure line (6), check valves prevent entry and discharge of fluid from steering lines. The piston in slave cylinder is manually moved by the steering shaft worm (3) and manual steering is accomplished by exchange of trapped fluid between slave cylinder (2) and all four ends of double pistons in steering motor (9).

TROUBLESHOOTING

All Early Models

13. Check fluid level before proceeding.

NO POWER STEERING

If fluid reservoir is full, operate another function, such as the rockshaft or a remote control cylinder to determine if the problem is lack of system pressure. If the other function also fails to operate, refer to paragraph 276 HYDRAULIC SECTION for possible causes.

NO OR POOR POWER STEERING TO LEFT

Leaking right pressure or right return valve

Steering motor piston seal failure

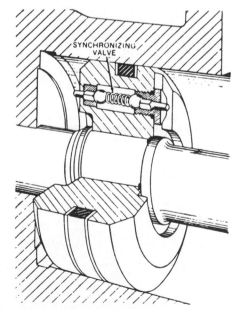

Fig. 17—Steering valve piston must be synchronized with steering motor for full turning action. Synchronization is automatically accomplished. When control valve piston reaches end of its stroke, the extended rod unseats the ball check valve allowing pressurized fluid to flow through piston until motor and valve are synchronized.

Fig. 15—Schematic view of typical early power steering system showing operating parts. Early 4630 models are similar, but use 2 operating pistons (Item 7). Refer to paragraph 11 for description.

1. Steering shaft
2. Slave cylinder
3. Actuating screw
4. Actuating collar
5. Operating levers
6. Pressure line
7. Operating piston
8. Steering spindle
9. Steering motor

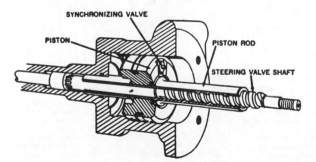

Fig. 16—Cross-sectional view of early steering valve operating piston, cylinder and steering shaft. Piston is moved up or down in cylinder by helical thread on steering shaft. The synchronizing valve which corrects for internal leaks is shown in Fig. 17.

NO OR POOR POWER STEERING TO RIGHT

Upper or lower steering valve piston rod seal failure

Left return or left pressure valve leaking

Steering check valve piston seal failure

Steering motor piston seal failure

NO OR POOR MANUAL STEERING TO LEFT

Steering valve cylinder piston seal failure

Synchronizing valve failure

Lower steering valve piston rod seal failure

Inlet check valve seat or seal failure

Steering check valve ball seat failure

Steering check valve piston stuck

Steering motor piston seal failure

NO OR POOR MANUAL STEERING TO RIGHT

Upper steering valve piston rod failure

Steering valve cylinder piston seal failure

Synchronizing valve leaking

Steering check valve piston seal failure

Steering motor piston seal failure

STEERING WANDERS TO LEFT OR RIGHT

Upper steering piston rod seal

Steering valve cylinder piston seal

Steering valve operating shaft collar assembly loose

Left pressure and return valve leaking

Steering check valve piston seal failure

Improper valve adjustment

Steering motor piston seal failure

FREQUENT SYNCHRONIZATION

Upper steering valve piston rod seal failure

Steering check valve piston seal failure

Steering motor piston seal failure

Synchronizing valve failure

EXCESSIVE STEERING WHEEL FREE PLAY

Steering valve shaft nut and rod loose

Collar loose

Air in steering system

BLEEDING

All Early Models

14. To bleed the system first allow engine to stop for at least one hour, remove the cowling and the small machine screw from bleed screw (Fig. 18). Attach a transparent hose to bleed screw and run free end back to reservoir. Start engine and run at slow idle speed. Turn steering wheel to full right, then to full left, to allow steering valve to synchronize with front wheels.

Leave steering wheel and front wheels in full left turn position, loosen bleed valve lock nut and back out bleed valve approximately 1/2-turn. With engine at slow idle and without moving front wheels, turn steering wheel very slowly to full right so that front wheels do not turn. Close bleed valve and allow front wheels to turn full right.

Repeat the procedure, if necessary, until air-free fluid is being returned to reservoir.

PRESSURE CONTROL (PRIORITY) VALVE

All Early Models

15. **OPERATION.** The Pressure Control (Priority) valve cuts off hydraulic flow to hydraulic lift system whenever system pressure drops below 1650-1700 psi, thus, giving priority to steering and brake units on all tractors, and the

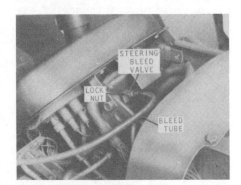

Fig. 18—To bleed the steering system, remove the cowl and attach bleed line to bleed screw then refer to paragraph 14 for procedure.

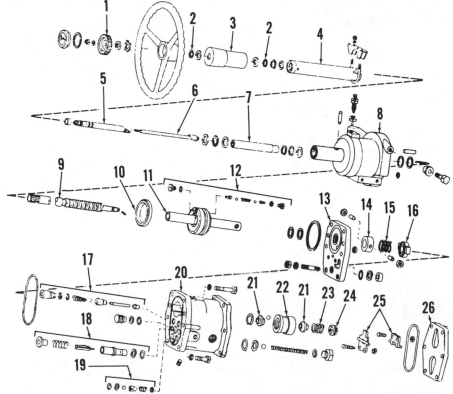

Fig. 20—Exploded view of early steering valve housing and associated parts. Telescoping model shown.

1. Release knob	8. Steering cylinder	14. Collar w/pins & rollers
2. Snap ring	9. Valve operating shaft	15. Spring
3. Sleeve	10. Seal	16. Nut
4. Tube	11. Piston & rod assy.	17. Control valve
5. Steering shaft	12. Synchronizing valve	18. Check valve
6. Release rod	13. Cylinder cover	19. Inlet check valve
7. Coupler		

20. Valve housing
21. Bearing race (2)
22. Operating collar
23. Spring
24. Slotted nut
25. Operating levers
26. Housing cover

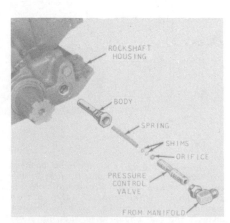

Fig. 19—Exploded view of pressure control (priority) valve showing component parts.

differential lock, if equipped. The priority valve is located in a bore in the rockshaft housing unless not so equipped, the valve is located in transmission cover. Refer to Fig. 19 for an exploded view.

In later models, the orifice became an integral part of the control valve. Relieve hydraulic pressure before testing or removing valve or pressure lines.

STEERING CONTROL UNIT

All Early Models

16. REMOVE AND REINSTALL. To remove the steering control unit, remove steering wheel, cowl, and hood.

Disconnect electrical connections to instruments, disconnect Speed-Hour Meter cable, then remove instrument panel.

Disconnect steering fluid lines, being sure to cap all fittings, then unbolt and remove the complete steering unit.

When reinstalling, bleed steering system as outlined in paragraph 14. Tighten steering wheel nut to a torque of 50 ft.-lbs.

16A. OVERHAUL. To disassemble the removed steering control unit, first remove lower cover (26—Fig. 20).

Check valve adjustment prior to complete disassembly, especially if previous tests have indicated incorrect valve adjustment. (Refer to paragraph 17 for adjustment procedures.)

Remove cotter pin and nut (24) from lower end of steering shaft, then unbolt and withdraw valve housing (20) and operating collar (22). Do not remove collar (22) before removing housing to prevent damaging valve levers. Check valve assembly (18) may fall out as housing is removed.

Remove nut (16), spring (15) and collar (14). The two rollers and pins in collar must be removed before collar can be withdrawn. Temporarily install steering wheel and turn wheel counter-clockwise to force cylinder cover (13) from cylinder housing (8), then with-

Fig. 21—Remove steering check valve stop and spring as unit is separated.

draw cover, piston and lower steering shaft from cylinder housing as a unit. Upper steering shaft is retained in steering column by a snap ring. Refer to Fig. 20 for parts identification on telescoping steering. Remove shaft if service is required on oil seal, bushing, shaft or housing. Withdraw springs from operating valves in valve housing, remove valve balls and inspect balls and seat for line contact. Note: If inlet check valve spring or guide damage is noted, there has been excessive flow through the assembly. Check for foreign particles in a control valve orifice or for particles holding the control valve or control valve ball off its seat. Check for seat damage due to foreign material. Check synchronizing valve assembly for sticking. This can prevent the proper synchronizing between valve and motor. Renew any parts that are damaged or worn.

Clean all parts by washing in clean solvent and immerse all parts including "O" rings and back-up washers in clean hydraulic fluid before assembly.

When reassembling, make sure synchronizing valves in piston are at one o'clock position and rod holes are horizontal as shown in Fig. 22. Tighten spring loaded nut (16—Fig. 20) on operating shaft until a gap of approximately 5/16-inch exists between nut (16) and collar (14). This tension provides the friction which gives a feeling of stability to the steering effort. Lever plugs control the end play of pivot

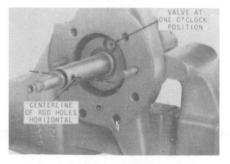

Fig. 22—Assemble piston with synchronizing valve at one o'clock and rod holes horizontal as shown.

Fig. 23—Install positioning clamps (JDH-3C) for valve adjustment.

shafts on operating levers (25). Adjust the plugs, if necessary, until levers turn freely but end play is limited to a maximum of 0.003 inch. The ball races (21) for operating collar (22) contain 13 loose bearing balls. Tighten nut (24) to a torque of 5 ft.-lbs., loosen to nearest castellation (if necessary) and install cotter pin. Operating collar (22) must turn by hand on shaft. Tighten the cap screws retaining valve housing to cylinder cover to a torque of 110 ft.-lbs., and the cylinder cover stud nuts to 55 ft.-lbs. Adjust the valve levers as in paragraph 17.

17. ADJUSTMENT. Install special neutral stops (JDH-3C) on housing as shown in Fig. 23. Use side of tools marked "3000-4000-5000." The tools will position lower edge of operating collar (E – Fig. 24) 0.030 inch from machined face of housing (H), which is neutral position. Turn steering wheel clockwise until operating collar contacts special tools, then attach a weight to steering wheel as shown in Fig. 23 to hold operating collar securely in this neutral position for steering valve adjustment.

Loosen locknuts on adjusting screws (A, B, C and D – Fig. 24) and make sure both operating levers have at least 0.003 inch free movement. Position a dial indicator on housing as shown in Fig. 23 so operating lever clearance can be checked at each adjusting screw. Adjust each screw to specified clearance using the following adjusting sequence: Adjust upper left return valve screw (A – Fig. 24) to provide 0.003 inch lever movement, then adjust upper right pressure valve screw (B) to obtain 0.001 inch movement. Adjust lower right return valve screw (C) to provide 0.003 inch lever movement, then adjust lower left pressure valve screw (D) to obtain 0.001 inch movement. This adjustment allows

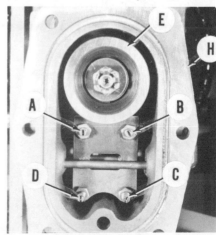

Fig. 24—Front view of valve with cover removed. Correct positioning of operating collar (E) with relation to housing face (H) is accomplished by the two blocks JDH-3C shown in Fig. 23.

pressure valves to open slightly before return valves.

Note that valve adjusting screw locations are as viewed from front side of housing with operating collar positioned at top as shown in Fig. 23. Also, adjustment can be more accurately made if operating lever shaft is pulled outward with light pressure to eliminate shaft free play as lever movement is being measured.

STEERING MOTOR

All Early Models

The steering motor and front support assembly can be repaired without removing from tractor. All early models except the 4630 have a single rack to drive the steering spindle. The 4630 model has a dual rack motor to drive the spindle from both sides at the same time (Figs. 25 and 26).

18. REMOVE AND REINSTALL. Should it become necessary to remove the steering motor, first remove hood and support front of tractor from a hoist by installing engine sling or other

suitable support. Remove fuel tank and air cleaner from front support. On tricycle models, remove wheels and pedestal assembly. On axle models, remove front axle and support as a unit if so desired.

Place a rolling floor jack or a hoist under steering motor to support the motor. Relieve hydraulic system pressure. Disconnect and plug fluid lines and remove any brackets and covers. Remove the cap screws securing steering motor to side frames and roll or hoist steering motor away from tractor.

To install the steering motor, reverse the removal procedure and bleed system as outlined in paragraph 14. Tightening torques are as follows:
Motor to side frames . . . . . . .275 ft.-lbs.
Steering spindle arm bolt . . .300 ft.-lbs.

19. OVERHAUL. If repairs are to be made with steering motor on the tractor, relieve hydraulic pressure in the system and disconnect all hydraulic lines to steering motor, including the motor-to-hydraulic pump bleed line. Drive wooden blocks between front axle housing and steering motor to prevent the motor from tipping to either side.

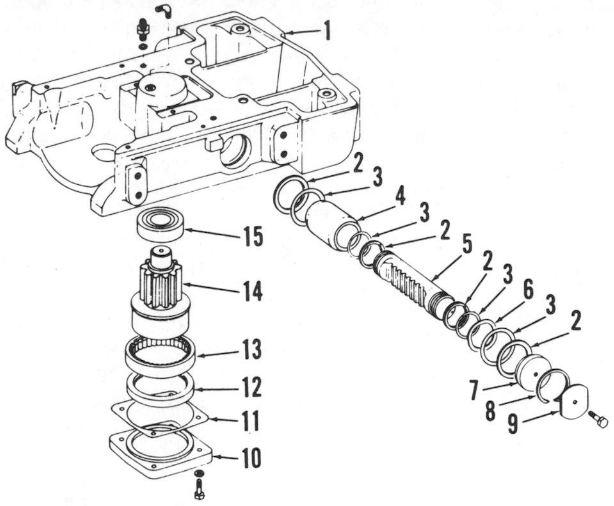

Fig. 25—Exploded view of typical steering motor used on all early models except 4630. (See Fig. 26). Model shown is convertible type (without pivot pin). Adjustable axle types are similar to Fig. 26, with only one piston.

1. Front support
2. Back-up ring
3. "O" ring
4. Sleeve
5. Piston
6. Spring washer
7. Plug
8. Snap ring
9. Washer
10. Quill
11. Gasket
12. Seal
13. Roller bearing
14. Spindle
15. Ball bearing

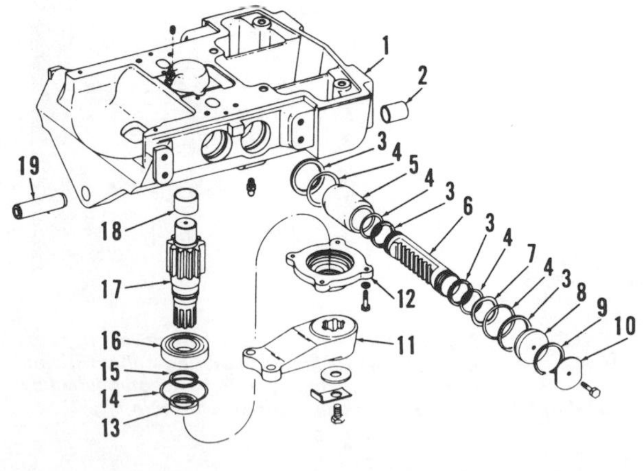

Fig. 26—Exploded view of early 4630 model steering motor. Items (3) through (10) have identical counterparts in rear bore, which also move spindle (17).

1. Front support
2. Bushing
3. Back-up ring
4. "O" ring
5. Sleeve (4)
6. Piston (2)
7. Spring washer
8. Plug (4)
9. Snap ring
10. Washer (cover)
11. Steering motor arm
12. Quill
13. Seal
14. "O" ring
15. Snap ring
16. Ball bearing
17. Spindle
18. Bushing
19. Pivot pin

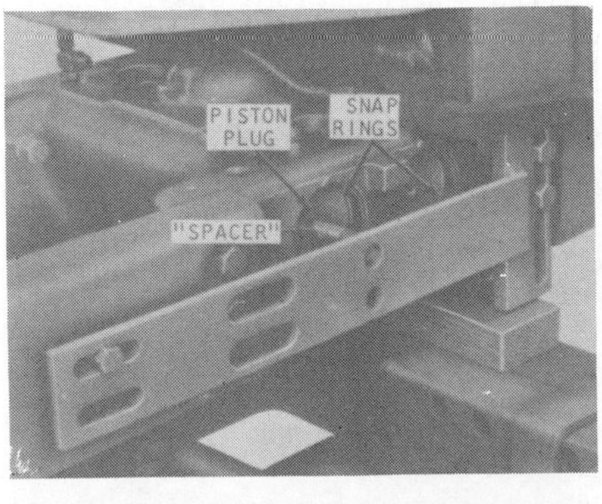

Place a floor jack under front axle assembly, and a moveable support under the steering motor and fuel tank to prevent forward tipping as assembly is moved. Remove all cap screws from steering motor and very carefully move the entire axle and steering motor assembly forward until rear set of holes line up with front holes in side frames. Reinstall 4 cap screws in steering motor at that point and tighten securely. Refer to Fig. 27. Remove piston plug covers. Install a bar to form a bridge across the piston plug to be removed, and insert a spacer to compress plug against spring washer (7—Fig. 26) so that snap ring (9) can be removed. Remove piston plug and spring washer. Remove spindle arm (11) and quill (12) and turn spindle to push piston out of sleeve. On 4630 models, make sure that proper snap rings (9) and plugs (8) are removed, since front and rear pistons will come out opposite sides of steering motor as spindle is turned. Remove spindle (17) and check all parts for wear or scoring. If sleeves (5) are scored or damaged, they may be removed and renewed. Piston O.D. should measure 2.6215 to 2.6235 inches, and sleeve I.D. should be 2.6250 to 2.6260 inches. If piston is renewed, sleeves should be renewed also. Bushing driver set No. 27797 may be used for sleeve removal. When removing the first sleeve, insert the bushing driver parts into inside of sleeve by going in through spindle hole. Handle may be installed through other sleeve and connected to driver. Second sleeve can be removed from opposite side after first sleeve is out. Check spindle bearings and bushings, and install new bearing with manufacturer's number toward outside. New sleeves should bottom in their bores, and upper spindle bushing should be installed until edge is flush to 0.015 inch below surface of spindle bore. Soak fiber back-up rings in hydraulic fluid for 30 minutes before assembly, and coat all parts

Fig. 27—Make bridge and use spacer to push in on piston plug for snap ring removal as shown.

with fluid. Install back-up rings toward center on piston, install "O" rings and insert piston into sleeve bores. Place spring washers (7—Fig. 26) against end of sleeve, and install piston plug "O" rings and back-up rings in plug bore with "O" rings to the inside of steering motor. Install piston plugs and compress plugs against spring washer, using the bridge and spacer as in disassembly.

Move piston all the way to one end of bore, and on 4630 models move pistons to opposite ends of each other. Install spindle into piston teeth, and make sure that the "V" mark on spindle lines up with right and left marks on housing when fully rotated each direction. (See Fig. 28). Install oil seal (13—Fig. 26) in quill with spring toward oil. On convertible steering motors, spindle bearing (13—Fig. 25) must be installed AFTER spindle is installed. Install bearing quill so that index marks match marks on housing. Install spindle arm on adjustable axle type motor with arm pointing to the rear when centered. Tighten cap screw to 300 ft.-lbs. torque, rap the arm hub several times and retorque. Bend lock plate into recess in arm.

Reinstall by reversing disassembly procedure and bleed system as outlined in paragraph 14.

STEERING SYSTEM (LATE TYPE)

Two different steering systems have been used. The late type steering covered in this section is used on the following tractors:

4030 Models . . . . After 008839
4230 Models . . . . After 019558
4430 Models . . . . After 027634
4630 Models . . . . After 010646

Some differences will be noted between certain of these late units, but all are equipped with full power steering. No mechanical linkage exists between the steering wheel and front unit; however, steering can be manually accomplished by hydraulic pressure without tractor normal hydraulic pressure.

Refer to paragraph 11 and following for services to earlier power steering system and components.

OPERATION

All Late Models

20. The power steering system consists of a metering pump (3—Fig. 30), a control valve (4) and a steering motor assembly (5 & 6). The power steering system, in normal operation, uses the tractor hydraulic system fluid which has been pressurized by the main hydraulic pump to turn the front wheels.

The metering pump (3—Fig. 30) is located at the lower end of the steering wheel shaft. The metering circuit is filled (2) with oil from the transmission and charging oil circuit. Turning the steering wheel and metering pump (3) causes the oil trapped in the metering circuit to move and apply pressure to one end of the control valve spool (4). The control valve spool will be moved

by the metering oil to one end of the valve body. When the control valve spool is not centered, a passage is opened allowing pressurized tractor hydraulic fluid to flow from inlet (1) through control valve and into one end of steering motor cylinder (5). During normal operation, oil from the feed back cylinders (6) is used to center the control valve spool after steering wheel is stopped or after turn is completed.

The steering system is equipped with check valves which trap oil in the system. In case of hydraulic system failure, the trapped oil is circulated within the steering system by the metering pump and pressure is exerted against both the steering motor piston and the feed back piston to turn the steering wheels.

Some differences will be noticed between control valves and steering motors used. For instance, the steering motor used on late 4630 model contains two operating pistons. The additional piston is identical to the one in the usual location, but is on the opposite (rear) side of the spindle.

TROUBLESHOOTING

All Late Models

21. Be sure that fluid is at correct level and that all hydraulic system filters are clean before suspecting the steering system. Operate another hy-

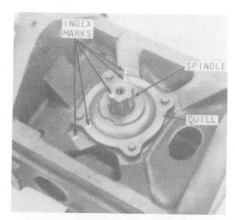

Fig. 28—V-mark on spindle should align with mark on housing at either end of piston stroke.

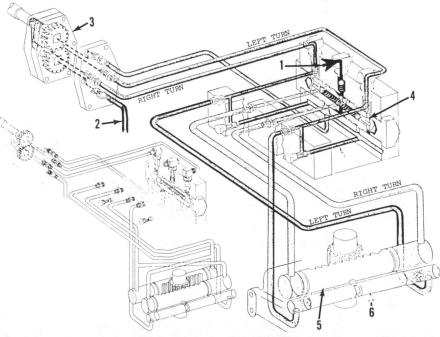

Fig. 30—Schematic view of steering system. The view at lower left is of type used on 4630 models.

draulic system function such as the rockshaft to determine whether system has pressure. Refer to paragraph 276 if hydraulic system fails to operate satisfactorily.

SLOW RESPONSE OR HARD STEERING

Tractor heavily weighted and/or not in motion.

Low hydraulic system pressure. Refer to paragraph 276.

Filter screens (17 & 23—Fig. 36) in control valve plugged or damaged.

Steering feed back orifices plugged, eroded or otherwise damaged. Remove filters (17—Fig. 36) and check orifices in valve body.

Cold oil in control circuit because of lack of circulation through metering pump.

Steering valve spool and/or body scored or damaged. Examine spool and bore of body for signs of sticking.

Metering pump leaking internally. Check pump for wear and/or damage.

Steering motor piston seal damaged. Check piston, seal and cylinder.

Feed back piston pin (40—Fig. 48) or spring damaged or stuck. Remove pin and check.

STEERING WHEEL CREEPS.
The steering wheel may require constant movement to maintain straight travel, may creep at locks and/or creep excessively when steering manually.

Steering motor feed back piston rings damaged.

Metering pump shims, wrinkled, folded or torn.

FRONT WHEELS LOCK TO ONE SIDE

The steering valve manual steering check valve seat (3—Fig. 36), ball (5) or spring (6) damaged. Inspect and install new parts as necessary.

Steering valve spool and/or body scored or damaged.

STEERING WHEEL CONTINUES TO TURN WITH WHEELS IN LOCK POSITION

Check make up valve (12 and 13—

Fig. 36) for damage. Install new parts as necessary.

STEERING WANDERS

Steering motor feedback piston pin (40—Fig. 48) stuck or spring (39) broken. Remove pin and inspect.

FRONT WHEELS TURN SHARPER IN ONE DIRECTION THAN THE OTHER

Steering motor spindle incorrectly indexed with front support. Refer to paragraph 28 and Fig. 49.

ERRATIC STEERING EFFORT

Metering pump return oil check valve in lower fitting (34L—Fig. 33) does not seat. Check and renew if damaged.

Steering control valve manual steering valves (3 and 5—Fig. 36) leaking. Install new parts if damaged.

FRONT WHEELS TWITCH WHEN ENGINE IS STARTED OR JERKS WHEN TURNING

Air in steering system. Check for loose or damaged connections and bores.

NO STEERING FEEL

Metering pump friction spring (32—Fig. 33) damaged. Install new spring.

METERING PUMP HOSE FAILURE

Return oil check valve in lower fitting (34L—Fig. 33) missing or damaged. Install new valve. Damage to check valve could be caused by pump friction spring (32) damage.

OIL LEAKAGE IN CONTROL SUPPORT OR OUT OF STEERING COLUMN

Metering pump oil seal (27—Fig. 33) damaged. Install new seal.

Metering pump shims torn, wrinkled or folded. Install new shims.

LEAKAGE THROUGH STEERING MOTOR BLEED LINE

Leakage past piston seal. Install new seals.

STEERING WHEEL MOVES UP AND DOWN WITH ENGINE RUNNING

Return oil check valve in lower fitting (34—Fig. 33) leaking. Install new parts as necessary.

LOSS OF MANUAL STEERING

Return oil check valve in lower fitting (34L—Fig. 33) of metering pump leaking or missing.

Metering pump leaking internally. Inspect and repair or renew as necessary.

Steering valve inlet check valve (19, 20, 21 and 22—Fig. 36) damaged.

Manual steering valves (3, 5 and 6—Fig. 36) leaking.

Make up valves (12 and 13—Fig. 36) leaking.

BLEEDING

All Late Models

22. The power steering will normally not require bleeding. If air is present in the power steering hydraulic system after service, cycling the steering wheel several times should purge air from lines and components. Air in operating system can be detected by twitch or jerk while turning, especially after just starting engine. Cycling the steering wheel several times may remove air, but check all lines and fittings for air and hydraulic leaks. Repair may be necessary to prevent reoccurrence.

PRESSURE CONTROL (PRIORITY) VALVE

23. The pressure control valve stops hydraulic fluid flow to the tractor hydraulic lift system when system pressure drops below 1600-1700 psi, thus giving priority to the steering and brake units. The pressure control valve is located in a bore of rockshaft housing.

To check operation, clean the area around the junction block (J—Fig. 31) and around the pressure control valve (V). Attach a 0-5000 psi pressure gage to the junction block and attach a hose to the selective control valve breakaway couplings so that oil can be circulated through the remote system. Move rate of operation lever, located just

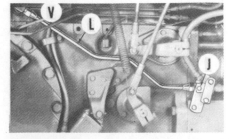

Fig. 31—View of tractor showing location of junction block (J), line (L) and pressure control valve (V).

Fig. 32—Exploded drawing of pressure control valve. Inset shows cross-section of retaining plug and elbow assembled.

J. Junction block
1. Elbow
2. Back-up ring
3. "O" ring
4. Snap ring
5. Plug
6. "O" ring
7. Pressure control valve
8. Shims
9. Spring
10. "O" ring (same as 12)
11. Valve body
12. "O" ring (same as 10)

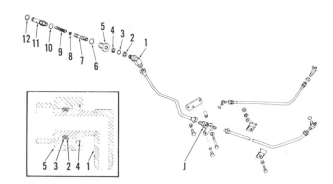

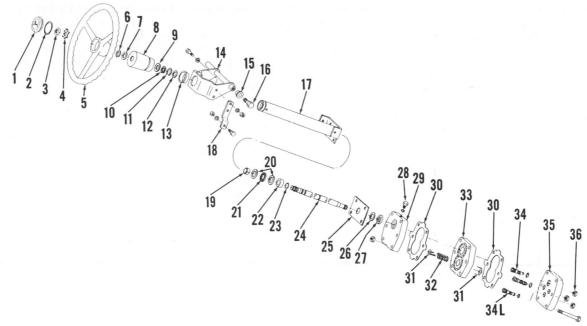

Fig. 33—Exploded view of metering pump and steering column used on some models. Refer to Fig. 34 for exploded view of telescoping steering column used.

1. Emblem	13. Tube collar	20. Thrust races
2. "O" ring	14. Pump support	21. Thrust bearing
3. Nut	15. Bushing	22. Collar
4. Lockwasher	16. Cap screw	23. Snap ring
5. Steering wheel	17. Steering column	24. Steering shaft
6. Snap ring (same as 10)	18. Strap	25. Plate
7. Washer (same as 9)	19. Bushing	26. Washer
8. Sleeve		27. Oil seal
9. Washer (same as 7)		28. Plug
10. Snap ring (same as 6)		29. Metering pump base
11. Snap ring		30. Shim
12. Oil seal		31. Friction plates
		32. Spring
		33. Pump body and gears
		34. Special hose connectors (not all the same)
		35. Pump cover
		36. Special nut

above breakaway couplings, to maximum (clockwise) position. Start engine and operate at 800 rpm then move the selective control valve lever to circulate hydraulic fluid through that system. The pressure indicated by gage attached to the junction block should be within range of 1600-1700 psi.

If pressure is incorrect, disconnect line (L—Fig. 31) from junction block (J) and control valve (V). Remove adapter (5—Fig. 32), then withdraw valve (7), shims (8) and spring (9). Add or deduct shims (8) as necessary to change pressure to within limits. Each shim represents about 50 psi. Reassemble and recheck pressure. Inability to adjust the pressure could be caused by damaged valve (7) and/or body (11), restriction in pressure or return passages in rockshaft housing or restriction in selective control valve circuit.

Spring (9—Fig. 32) should exert 45-55 lbs. when compressed to height of 3½ inches.

METERING PUMP AND STEERING COLUMN

All Late Models

24. Some repairs can be accomplished without removing the steering column and metering pump from tractor. Use extreme care to be sure that dirt is not permitted to enter the steering system

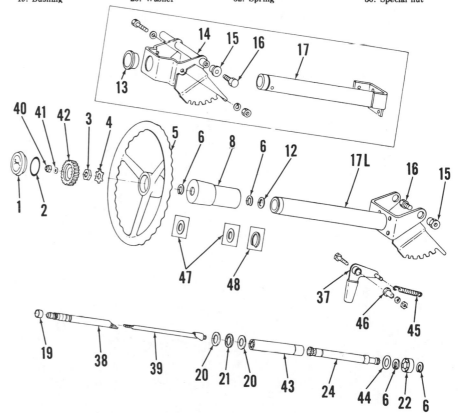

Fig. 34—Exploded view of telescoping steering column used on some tractors. The parts shown in boxes are earlier type. Refer to Fig. 33 for illustration of metering pump and for legend except the following.

17L. Later welded column and support assembly	39. Column release rod	43. Steering shaft coupling
38. Upper steering shaft	40. Special jam nut	44. Washer
	41. Special washer	45. Spring
	42. Release knob	46. Spacer
		47. Washers (early)
		48. Snap ring (early)

lines or ports.

Disconnect battery ground cables, relieve hydraulic system pressure and remove steering wheel. Remove the left and right cowl panels and the rear control support cover. Remove snap ring from upper end of steering shaft and withdraw the sleeve from around top of steering shaft. Disconnect the four hydraulic lines from metering pump and cover all openings. Detach spring from lug on steering column, remove the two pivot screws, then tilt the metering pump and steering column up and out toward front of tractor.

Both fixed (rigid) and telescoping steering columns have been used. Refer to Fig. 33, Fig. 34 and Fig. 35. The metering pump should be unbolted and separated from lower end of steering column before servicing either assembly.

To disassemble the removed metering pump, loosen the two remaining screws, break the seal between pump body and pump cover then remove screws gradually. Examine parts of pump carefully, especially the two 0.0005 inch thick shims (30—Fig. 33). Internal or external leakage can be caused by wrinkled, torn or otherwise damaged shims. The special fittings (34) can be removed after removing special nuts (36). Be sure to reinstall fittings in proper ports.

Free length of friction spring (32) should be 0.735 in. and the spring should exert 81-99 lbs. force when compressed to height of 0.52 inch. Locate JDH-42-2 or equivalent dowels in the two center holes of cover (35) and four cap screws in corner holes to align pump while assembling. Carefully install one shim (30) over cover (35), then

install body (33) onto cover and shim. Position gears in body bores. Make sure that marks on gears are up and that splined top gear is in correct (upper) bore. Install friction spring (32) and plates (31) in idler gear. Carefully position the second shim (30) onto pump body and lower pump base (29) over dowels onto remainder of pump. Use care to not damage friction plate or shims when positioning the pump base. Install nuts on the two lower retaining screws and tighten nuts to 35 ft.-lbs. torque, then remove alignment dowels.

Assemble remaining parts and attach to column. Tighten the four remaining screws to 35 ft.-lbs. torque. Check unit for leaks and proper operation after reinstalling and bleeding (cycling) the system.

STEERING VALVE

All Late Models

25. The control circuit is operated by the metering pump which moves the control valve which in turn directs pressurized hydraulic fluid to the steering motor cylinders to move the front wheels. The actual location of the control valve will depend upon specific model of tractor. It will not be necessary to remove the valve for some service such as cleaning filters. The valve spool (7—Fig. 36) and body (24) are available only as a matched set. Specific service will depend upon difficulties encountered. Be sure that all parts are clean when assembling.

STEERING MOTOR

All Late Models

The steering motor is contained in the front support assembly and most repair can be accomplished without removing the front support from the tractor. Some differences may be noticed between standard two wheel drive tractors and models with power front wheel drive. Model 4630 tractors use two racks in steering motor to turn the spindle. Where applicable, differences will be noted.

26. **STEERING MOTOR ACCESS.** Most service to the steering motor can be accomplished after sliding the front support forward in the frame rails.

Support tractor at sides of frame and support the fuel tank and steering motor (front support housing). Remove muffler, air intake tube, hood side shields, side grille screens and hood. Relieve hydraulic pressure and dis-

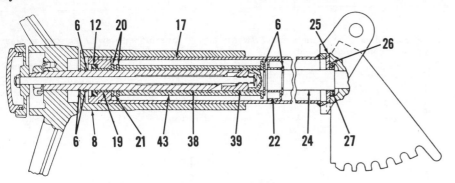

Fig. 35—Cross-section of telescopic steering column. Refer to Fig. 34 for exploded view.

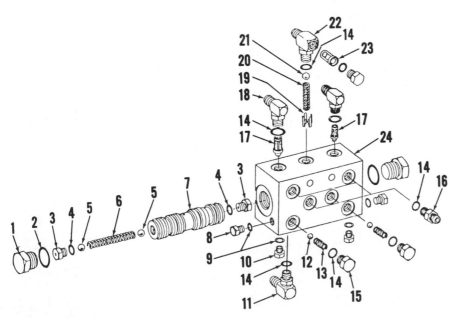

Fig. 36—Exploded view of typical power steering control valve. The additional parts shown in Fig. 37 are used on some models with power front wheel drive.

1. Plugs	7. Control valve spool	14. "O" rings	20. Spring
2. "O" rings	8. Plug	15. Plug	21. Inlet check valve
3. Seats	9. "O" rings	16. Connector fittings	ball (3/8-inch steel)
4. "O" rings	10. Plugs	17. Control circuit	22. Pressure inlet
5. Steel balls (5/16-in.)	11. Elbow	filters	fitting
for manual steering	12. Steel balls (3/8-inch)	18. Elbows	23. Inlet filter
6. Spring	for make up valves	19. Guide	24. Valve body
	13. Springs		

connect battery ground cables. Remove fuel tank leak-off line, shut-off fuel valve at bottom of tank and disconnect steering motor bleed line from connector above the steering motor. Remove fuel supply line from between fuel tank and fuel filters. Detach the four steering lines from steering control valve and cover all of the openings and ends of lines. Remove oil cooler return hose and clamps securing hose to fuel tank.

Fig. 37—View of additional parts installed in power steering control valve of some tractors with power front wheel drive.

25. Special plugs
26. "O" rings
27. Shims
28. Springs
29. Guides
30. Steel balls same as (5—Fig. 36)
31. Valve seats
32. "O" rings
33. "O" rings
34. Valve seats
35. Steel balls (3/16-inch)
36. Guides
37. Springs
38. Shims
39. "O" ring
40. Special plugs

Fig. 40—View of late 4630 tractor with front support (1) moved forward in frame rails (F) and held in position by screws (S). The piston plugs (37) and other components of the steering motor are accessible for service. Other models are similar.

Disconnect radiator support rod and the hydraulic line clamp from radiator. Disconnect wire from fuel gage sending unit in tank, also disconnect horn and air conditioner control wires if so equipped. Remove lower frame plate or hydraulic line guard. Remove radiator mounting screws and the upper frame plates over the main hydraulic pump. Loosen the baffle plate. Attach hoist to front support using special lift bracket (JDG-3) or equivalent. Use frame side stands or equivalent to securely hold rear of tractor while removing front support (steering motor). Remove screws which attach the front support to the frame side rails, then carefully slide the front

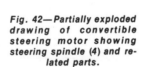

Fig. 41—View of bar (B) installed and spacers (S) being used to push piston plugs (35) in so that snap ring (36) can be removed. Procedure is similar for late 4630 model.

Fig. 42—Partially exploded drawing of convertible steering motor showing steering spindle (4) and related parts.

1. Front support
2. Axle pivot bracket
3. Steering arm
4. Spindle
5. Single wheel tricycle fork
6. Dual wheel tricycle pedestal
7. Roll-O-Matic pedestal
13. Pivot pin
14. Bushing
20. Ball bearing
21. Roller bearing
22. Spacer (some models)
23. Oil seal
24. Gasket
25. Steering motor spindle quill

support forward until two bolts can be installed through front holes in frame rails and into the rear threaded holes in support as shown at (S-Fig. 40).

Most repair can be accomplished as outlined in paragraph 28 with the front support attached to the tractor.

Tighten screws attaching frame rails to steering motor to 275 ft.-lbs. Bleed system after installation is complete by turning steering wheel from lock to lock several times. Check for leaks and proper operation.

27. **REMOVE AND REINSTALL.** To remove the complete steering motor and front support assembly, proceed as follows: Remove hood and support front of tractor from a hoist by installing engine sling or other suitable support. Remove fuel tank and air cleaner from front support. On tricycle models, remove wheels and pedestal assembly. On axle models, remove front axle and support as a unit if so desired.

Place a rolling floor jack or a hoist under steering motor to support the motor. Relieve hydraulic system pressure. Disconnect and plug fluid lines and remove any brackets and covers. Remove the cap screws securing steering motor to side frames and roll or hoist steering motor away from tractor.

To install the steering motor, reverse

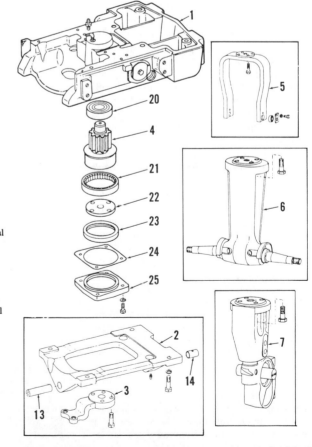

the removal procedure then bleed system by turning the steering wheel to each extreme several times. Tightening torques are as follows:

Steering motor to
to side frames .... 275 ft.-lbs.
Steering spindle arm bolt ... 300 ft.-lbs.

28. **OVERHAUL.** The following overhaul procedure can be accomplished with the steering motor and front support moved forward in the frame rails as outlined in paragraph 26 or with unit removed from tractor as outlined in paragraph 27.

Attach a bar (B—Fig. 41) between frame mounting holes and remove the piston caps (cover washers) if used. Use a spacer (S) to push piston plugs (35) in, then remove the retaining snap rings (36). Remove piston plugs, back-up rings, "O" rings and spring washers.

On tricycle models, unbolt and remove the tricycle pedestal (6 or 7—Fig. 42) or yoke (5) from spindle (4). On axle models with convertible steering gear, detach tie rods from steering arm (3) then unbolt and remove steering arm from spindle (4). On standard axle steering motor, detach tie rods from steering arm (3—Fig. 44 or Fig. 45), remove screw (18) then pull steering arm (3) from spindle (4).

Turn the spindle to one extreme (either left or right), then unbolt quill (25—Fig. 43, 15—Fig. 44 or 15—Fig. 45). The quill, seal, steering spindle and lower bearing can now be removed. Align the feed back piston pin (40—Fig. 43, Fig. 44 or Fig. 45) with notch in housing as shown in Fig. 46, then turn piston so that pin and spring can be removed. A special tool (T—Fig.46) is available, which engages the two depressions in top of piston to facilitate turning the piston. Withdraw steering motor pistons (31—Fig. 43, Fig. 44 or Fig. 45). Remove snap rings (47) then withdraw plugs (46) and feed back piston (43).

Check pistons (31) and sleeves (28) for wear or scoring. Sleeves can be pulled from housing using appropriate sized piloted puller. Examine parts against the following values.

Piston (31) O.D. .... 2.621-2.623 in.
Sleeve (28) I.D. .... 2.625-2.626 in.
Pin (40) O.D. .... 0.7357-0.7363 in.
Bore for pin (40) in
piston (31) .... 0.7365-0.7375 in.
Pin (40) length .... 2.58 in.
Spring (39) pressure at
0.74 in. .... 45-55 lbs.
Feed back piston (43)
O.D. .... 1.499-1.500 in.
Bushing (41) I.D. .... 1.501-1.506 in.
Bushing (41) O.D. ..... 1.8515-1.8525 in.

Feed back bushings (41) should only be removed if new bushings are to be

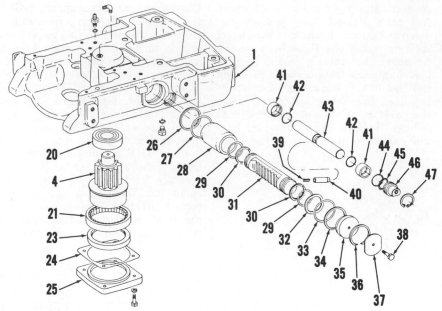

Fig. 43—Exploded view of convertible steering motor also shown in Fig. 42. Refer to Fig. 42 for legend except the following.

26. Back-up rings	32. Spring washers	37. Washers	42. "O" rings
27. "O" rings	33. "O" rings	38. Retaining screws	43. Feedback piston
28. Cylinder sleeves	34. Back-up rings	39. Spring	44. "O" rings
29. "O" rings	35. Piston plugs	40. Special pin	45. Back-up rings
30. Back-up rings	36. Snap rings	41. Sleeves	46. Plugs
31. Piston			47. Snap rings

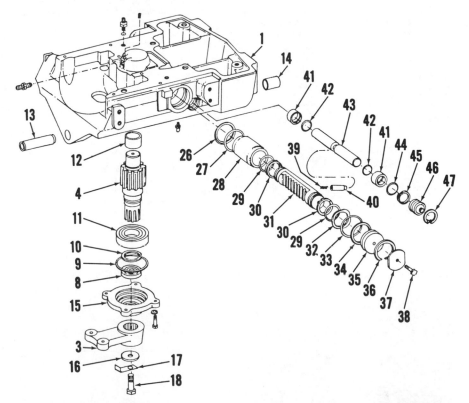

Fig. 44 — Exploded view of standard front support and steering motor used on late 4030, late 4230, and late 4430 models without convertible steering motor. The unit shown above cannot be used with tricycle front wheel or wheels. Refer to Fig. 43 for legend except the following.

1. Front support		15. Bearing quill
3. Center steering arm	9. "O" ring	12. Bushing	16. Washer
4. Spindle	10. Snap ring	13. Pivot pin	17. Lock plate
8. Oil seal	11. Ball bearing	14. Bushing	18. Cap screw

If sleeves (28—Fig. 43, Fig. 44 or Fig. 45) were removed, install fiber back-up ring (26) in groove, then install "O" ring (27) in same groove, but toward outside of back-up ring. Use a piloted driver and carefully drive sleeves into position until bottomed in casting bore. Repeat procedure for remaining sleeves. A total of four sleeves are used on each 4630 tractor. Two sleeves are used on other models.

Install "O" rings (42), then slide feed back piston (43) into bore. Center the feed back piston so that pin groove is aligned with notch (N—Fig. 46) in spindle bore.

Install back-up rings (30—Fig. 43, Fig. 44 or Fig. 45) and "O" rings (29) in grooves of piston (31). The back-up rings should be toward center of piston and the concave (hollowed) side of back-up ring should be against "O" ring. Slide the steering motor piston (31) into cylinder, with the two counter sunk holes toward right side. Turn the piston to align hole in piston with notch (N—Fig. 46), grease the spring (39—Fig. 43, Fig. 44 or Fig. 45) and pin (40), then slide the spring and pin into hole in piston. Turn the steering motor piston (31) to position the pin (40) in groove of feed back piston (43). On 4630 model, install the rear piston using similar procedure.

Install back-up ring (34) and "O" ring (33) in groove, position spring washer (32) then slide plug (35) in bore. Use a bridge and spacer as shown in Fig. 41 to compress spring washer enough to install retaining snap ring (36). Install back-up rings, "O" rings, spring washer and plugs (33, 34, 35 & 36—Fig. 43, Fig. 44 or Fig. 45) in the remaining cylinder bores.

Install oil seal (8—Fig. 44 and 45) with spring loaded lip up toward spindle gear (4). Push the (front on 4630 model) steering motor piston (31) to the right extreme of travel. On 4630 models, push the rear steering motor piston to the extreme left to travel. On all models,

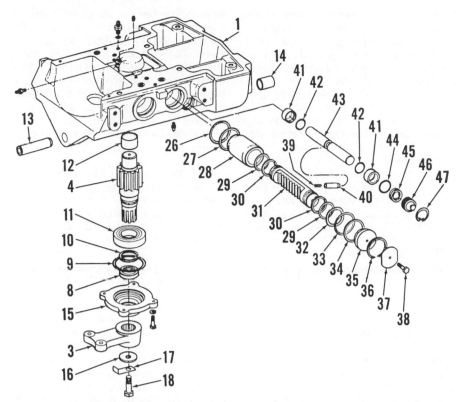

Fig. 45—Exploded view of the front support and steering motor used on late 4630 model. Two pistons (31) are used and are located in bores located in front of and behind the spindle (4). Refer to Fig. 43 and Fig. 44 for legend and to Fig. 48 for cross-section of this unit.

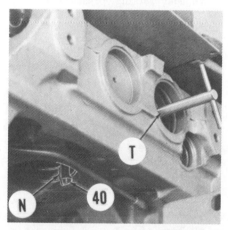

Fig. 46—A special tool (T) is available for turning the front piston permitting pin (40) to be removed through notch (N).

Fig. 47—View showing installation of plug (46) and snap ring (47) in one end of feed-back piston bore.

installed. Use a blind hole puller to remove bushings (41) and be careful not to catch lip of casting while attempting to remove. See Fig. 48. Press new bushings (41) in until seated against inner shoulder in bore.

Check condition of bushing (12—Fig. 44 and Fig. 45). Excessive wear could be caused by leakage or inadequate lubrication. Drive new bushing (12) into bore until flush to 0.015 inch below flush with casting bore. Identification numbers on bearing (11) should be toward outside (down).

Dip all parts in John Deere Hy-Guard Transmission oil or equivalent while assembling. Soak fiber back-up rings in oil for 30 minutes before assembling.

Fig. 48—Cross-section of steering motor typical of late 4630 model. Refer to Fig. 43 and Fig. 44 for legend. Other late models are similar except that rear piston and associated parts are not used.

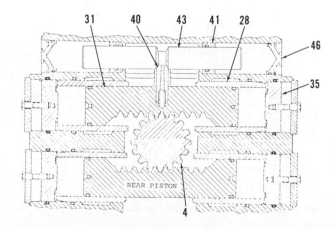

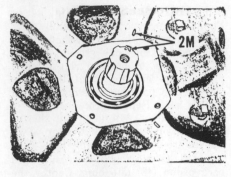

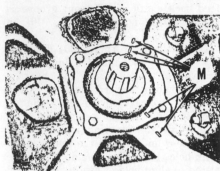

Fig. 49—Align the two marks (2M) when installing spindle. Align marks on bearing quill as shown at (M). Refer to text.

slide the spindle into position with "V" mark on spindle aligned with "I" mark at right front as shown at (2M – Fig. 49). Install "O" ring (9 – Fig. 44 or Fig. 45) in groove, then install quill (15) with mark on quill aligned with other two marks as shown at (M – Fig. 49). Assemble models with convertible steering motor (Fig. 43) using a similar procedure.

POWER FRONT WHEEL DRIVE

All models may be equipped with a Power Front Wheel Drive unit. The front wheels are driven by hydraulic fluid supplied by the tractor main hydraulic system.

OPERATION

All Models So Equipped

30. The hydraulic power Front Wheel Drive unit consists of an axial piston hydraulic motor in each front wheel hub which turns the wheel through a planetary gear reduction unit.

Power is supplied by the constant pressure, variable volume Main Hydraulic System Pump, thus providing a specified amount of torque to front wheels without regard to tractor ground speed. This design permits a

"Power Assist" traction boost which need not be precisely synchronized with tractor drive gears, and which eliminates possibility of damage or overload due to wheel slippage, altered tire size or tread wear.

The control valve is electrically operated by switches contained in Shift Lever Quadrant, and automatically coupled to the directional control solenoid. On Syncro Range and Power Shift tractors, power front drive unit is automatically disengaged above 6th gear in "Low Torque" range, above 4th gear in "High Torque" range; or whenever Inching Pedal (Power Shift Models) or Clutch Pedal (Syncro Range and Quad Range Models) is depressed.

On Quad Range models "High Torque" is limited to the 4 speeds in "A" range and 1st speed in "B" range. When speed selector lever is placed in 2nd or 3rd in "B" range, and 1st, 2nd or 3rd in "C" range, the high torque is automatically switched to low torque drive if operating switch is in the engaged position. Fourth gear disengages front wheel drive unit in all but "A" range in both high and low torque positions, and no front wheel drive is possible in "D" range.

The panel mounted operating (selector) switch has three positions: Off, High Torque and Low Torque.

With operating switch in Off position, the wheels do not drive, but are allowed to rotate freely.

When the operating switch is in High Torque position, oil pressure from the main hydraulic pump is directed to the drive motors in both front wheel motors and to the front wheel drive brake.

With the operating switch in Low Torque position, the control valve

directs pressurized oil to only one of the front wheel drive motors. Oil returning from this front wheel drive motor flows back to the control valve and is directed to the other front wheel drive motor. The drive motors in Lov torque drive are connected hydraulically in series and since both drive wheels displace the same volume, both wheels will rotate a similar amount with a minimum amount of slippage.

31. **CONTROL VALVE.** Refer to Fig. 52 for a cross-sectional view of front wheel control valve. The unit contains three solenoid-operated pilot valves; F (Forward), R (Reverse) and T (Torque), and three hydraulically actuated operating valves; (Pressure Control Valve) which gives priority to other hydraulic functions if necessary, (Direction Valve) and (Torque Valve). A pressure relief valve is mounted near pressure inlet.

The three pilot valves are spring loaded in the closed position and direction valve is spring centered, blocking fluid passage to wheel motors whenever electric power is cut off to control valve. When forward or reverse solenoid is electrically energized, the pilot valve opens and admits pressure to one end of direction valve, shifting the valve and routing pressure fluid to front wheel motors in forward or reverse direction depending on which way valve is moved.

Torque valve is spring loaded in High Torque position. When Torque Pilot Valve (T) is not energized, flow through the direction valve is delivered at equal pressure to each front wheel motor. In Forward position, when torque pilot valve is energized, pressure fluid enters piston end of torque valve and moves it against spring pres-

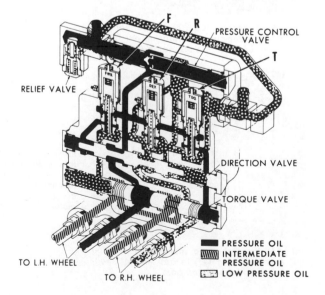

Fig. 52—Cross-sectional view of Power Front Wheel Drive control valve showing component parts. Shaded areas indicate pressurized operating fluid with valve in forward, low torque position.

■ PRESSURE OIL
▨ INTERMEDIATE PRESSURE OIL
▦ LOW PRESSURE OIL

sure to route full volume to flow to left wheel motor. Discharge port of left wheel motor is connected through control valve to inlet port of right wheel motor and discharge port of right wheel motor to discharge port of valve. The two wheel motors, being series connected, now deliver half the torque on half the volume of fluid.

In reverse position, when torque pilot valve is energized, pressure fluid enters piston end of torque valve and moves it against spring pressure to route full volume flow to right wheel motor. Discharge port of right wheel motor is connected through control valve to inlet port of left wheel motor and discharge port of left wheel motor to discharge port of valve.

Electrical switches control current to the solenoids of control valves. Refer to Fig. 53 for schematic.

32. WHEEL DRIVE UNIT. Refer to Fig. 54. The front wheel drive unit consists of a fixed displacement axial piston motor and primary sun gear (A), which drives a dual reduction planetary

gear unit (B). The primary planet carrier is splined to a secondary sun gear which drives hub (C).

The planetary secondary ring gear (D) floats in its housing and is coupled to spindle carrier by hydraulic pressure whenever power is applied. The wheel manifold contains check valves to admit pressure to brake pistons and wheel motor pistons simultaneously. One planet brake piston contains a relief valve to prevent damage due to pressure buildup during neutral operation. In forward direction, pressure oil is routed through the upper hoses to front wheel assembly, and return oil is directed through lower hoses to control valve and then to the reservoir. For reverse direction, pressure oil is directed to the lower hoses and return is through upper hoses. Refer to Fig. 55 for oil pressure flow through manifold. The upper check valve body has a 0.031 inch orifice which allows brake piston apply pressure to bleed to low pressure passage after controls are moved to the neutral, or off position. A relief valve is provided in the center of manifold to prevent damage due to excessive pressure buildup.

TROUBLESHOOTING

All Models So Equipped

CAUTION: If engine is to be run while checking Power Front Wheel Drive, shut off the hydraulic pump and disconnect the wiring harness near the solenoids at the electrical connectors. This will prevent tractor movement if a short circuit or accidental switch movement should attempt to engage the front wheel drive.

33. On all models, electrical circuits can be checked whenever key switch is turned on, but on models with Perma-Clutch or Power Shift the clutch pressure switch must be by-passed using a jumper wire. Solenoids close with an audible click. With key switch on and clutch switch jumper wire installed, test as follows:

(1) Move selector switch from "OFF" to "LOW TORQUE" (lower) position; torque solenoid should engage with an audible click.

(2) Move selector switch back to "OFF" position and Gear or Speed selector lever to 1st or 2nd forward

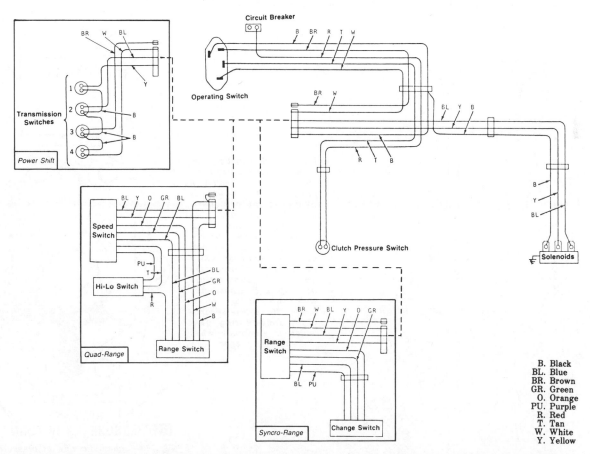

B.	Black
BL.	Blue
BR.	Brown
GR.	Green
O.	Orange
PU.	Purple
R.	Red
T.	Tan
W.	White
Y.	Yellow

Fig. 53—Wiring diagram for Power Front Wheel drive. The wiring and switches which are different with various transmissions are shown in the insets.

gear; then, move selector switch to "HIGH TORQUE" (upper) position. Forward solenoid should engage with an audible click.

(3) Repeat test 2 except with Gear or Speed selector lever in 1st or 2nd reverse gear. Reverse solenoid should engage with an audible click.

NOTE: HIGH TORQUE (upper) position of selector switch is used to test forward and reverse solenoids to keep from energizing torque solenoid.

If any of the solenoids fail to engage at the proper time, continuity checks of the circuits should be conducted to determine the cause. Refer to Figs. 53, 56, 57 or 58 for wiring diagrams of power front drive units. Check red wire from clutch pressure switch to circuit breaker and make sure circuit breaker is not open.

Internal hydraulic leakage or failure, because of the closed center system, will usually be signalled by heat or noise.

ELECTRICAL SYSTEM

All Models So Equipped

34. **ADJUSTMENT.** Except for circuit breaker terminal, terminals for solenoid connections and rotary switch connections, quick disconnect terminals are used.

On Syncro Range models, the two rotary switches must be synchronized as follows:

Remove console cover and be sure key is off, or remove from switch. Place shift lever in Fifth Gear.

Disconnect rod from Speed Range Switch and turn switch counter-clockwise against internal stop. Turn switch clockwise THREE detents, then adjust control rod yoke if necessary until rod can be reconnected to switch arm.

NOTE: Be sure all shift lever supports and switch brackets are tight before making the above adjustments.

On Quad-Range models, remove console cover and be sure key is off, or removed from switch. Place shift levers in C-1 position and disconnect yoke from Speed Switch arm (left lever). Turn speed switch counter-clockwise to internal stop. Turn clockwise THREE detents and adjust yoke, if necessary, so that clevis pin will line up with hole in switch arm, and reconnect yoke. Disconnect yoke from Range Switch arm (right lever) and turn switch counter-clockwise to internal stop. Turn clockwise THREE detents and adjust yoke if necessary, so that clevis pin will line up with hole in switch arm, and re-

connect yoke. Move both levers to all positions and check for proper adjustment.

On Power Shift models, disconnect the wiring harness from the 4 transmission switches and check for continuity as follows:

TOP switch (No. 1)—closed in all reverse speeds.

2nd switch—closed in all forward speeds.

3rd switch—closed in 1st through 6th speeds.

BOTTOM switch (No. 4)—closed in 1st

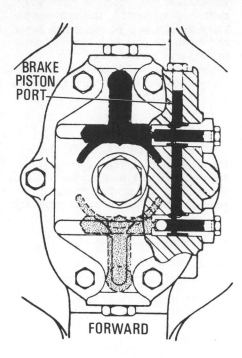

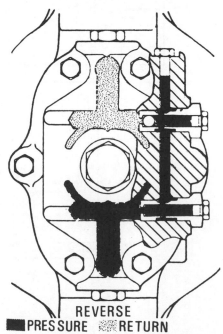

Fig. 55—Schematic view of pressure and return oil through wheel oil manifold. Brake piston port is pressurized in both directions.

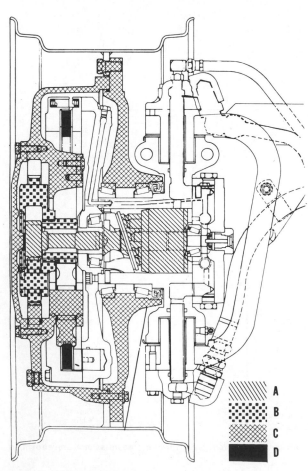

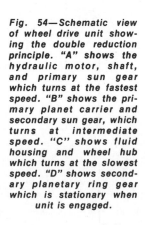

Fig. 54—Schematic view of wheel drive unit showing the double reduction principle. "A" shows the hydraulic motor, shaft, and primary sun gear which turns at the fastest speed. "B" shows the primary planet carrier and secondary sun gear, which turns at intermediate speed. "C" shows fluid housing and wheel hub which turns at the slowest speed. "D" shows secondary planetary ring gear which is stationary when unit is engaged.

through 4th speeds. Refer to Fig. 57 for wiring diagram. If adjustment is necessary, remove switches and add or remove shim washers to make switches close ONLY as outlined above.

The clutch pressure switch should be checked for continuity, and should make contact as the clutch engages.

OVERHAUL

All Models So Equipped

35. PILOT VALVES. Solenoid coil (2—Fig. 59) can be removed from pilot valve after removing nut (1), without disturbing hydraulic circuits. Pilot valves can be removed for inspection, renewal, or renewal of "O" rings after removing solenoid coil. Main hydraulic pump should be shut off and system pressure relieved before attempting removal of valve unit.

Valve spool and bushing are only available as an assembly (10). All other parts are available individually, and seals are provided in the control valve seal kit assembly.

36. CONTROL VALVE. Refer to Fig. 60 for an exploded view of control valve assembly. Valve spools (3, 10 and 15) or body (2) are not available separately but spools may be removed for inspection or renewal of other parts. All seals including those on pilot valves (Fig. 59) are available in a valve sealing kit. Shims (6—Fig. 60) control setting of priority valve (3). Operating pressure should be 1930-1970 psi with fluid at operating temperatures and engine at 2150 rpm.

Refer to Figs. 52 and 60 and assemble by reversing disassembly procedure, making sure valves are installed correct end forward. New sealing rings should always be used when assembling control valve unit.

36A. WHEEL MANIFOLD. Refer to Fig. 61 for an exploded view of wheel manifold and associated parts, and Fig. 55 for a schematic view of oil flow. Valves (7-12 and 13-17—Fig. 61) can be removed for inspection or service without removing manifold plate. Plate can be removed without taking weight off wheel; proceed as follows:

Drain wheel housing by turning plug (30—Fig. 68) downward and removing plug. Housing capacity is approximately 9 quarts. Turn wheel for access to manifold cap screws and loosen screws evenly. Wheel motor spring should push manifold approximately 1/8-inch away from motor housing (spindle) as screws are loosened. Valve plate (1—Fig. 61) and/or bearing plate

(13—Fig. 62) may remain with motor or be removed with manifold. Do not allow plates to drop as manifold is removed. Upper check valve body (17—Fig. 61) has an orifice, and is

longer than the lower, which has no orifice.

Install by reversing the removal procedure. Manifold plate should easily slide on attaching bolts to within 1/8-

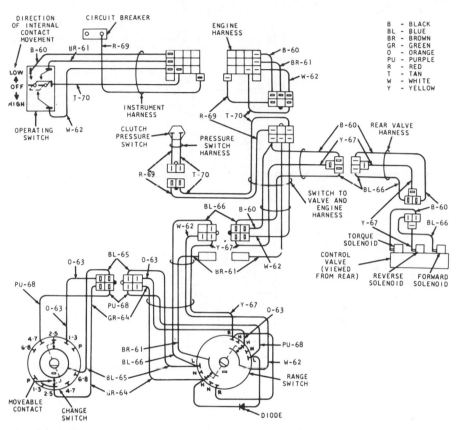

Fig. 56—Wiring diagram for Power Front Drive electrical system used on Syncro Range models.

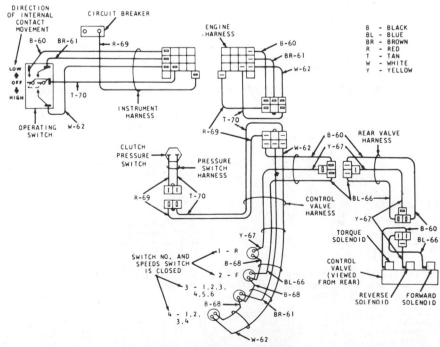

Fig. 57—Power Front Drive electrical system wiring diagram used on Power Shift Models.

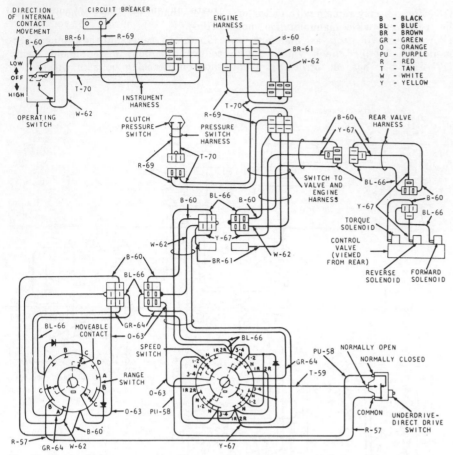

B - BLACK
BL - BLUE
BR - BROWN
GR - GREEN
O - ORANGE
PU - PURPLE
R - RED
T - TAN
W - WHITE
Y - YELLOW

Fig. 58—Power Front Drive electrical system Wiring Diagram used on Quad-Range Tractors.

37. WHEEL MOTOR. The hydrostatic wheel motor is axial piston type shown exploded in Fig. 62. Wheel motor can be removed as outlined in paragraph 38 and overhauled as in paragraph 39.

38. REMOVE AND REINSTALL. To remove the hydrostatic wheel motor, first remove manifold as outlined in paragraph 36A and primary planetary unit as in paragraph 40. If brass bearing plate (13—Fig. 62) remained with motor, insert a suitable wood dowel in one of the small round drain holes to use as a pry. NEVER use a screwdriver or steel wedge of any kind.

Grasp inner bearing cone (3—Fig. 61) and withdraw wheel motor (2—Fig. 68), swash plate (3), motor shaft (7) and outer bearing cone (8) as a unit from motor housing. If swash plate binds in housing or drags on locating pin, pressure can be applied at outer end by inserting a wood dowel or brass drift in spline end of shaft (7).

NOTE: Do not rotate assembly when out of housing, as damage may occur to motor parts while not supported by housing.

Install by reversing the removal procedure, making sure thick side of swash plate ramp is toward rear of tractor and that locating dowel pin enters appropriate notch in swash plate.

39. OVERHAUL. To overhaul the removed wheel motor unit, use specially cut pieces of clean cardboard, sheet gasket material or plastic to protect the polished inner face of cylinder

inch of motor housing. Considerable pressure is required to compress motor spring, but compression should be even without binding. Tighten manifold cap screws to a torque of 35 ft.-lbs.

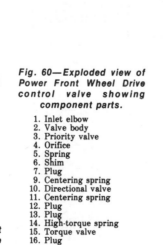

Fig. 60—Exploded view of Power Front Wheel Drive control valve showing component parts.

1. Inlet elbow
2. Valve body
3. Priority valve
4. Orifice
5. Spring
6. Shim
7. Plug
8. Centering spring
9. Centering spring
10. Directional valve
11. Centering spring
12. Plug
13. Plug
14. High-torque spring
15. Torque valve
16. Plug

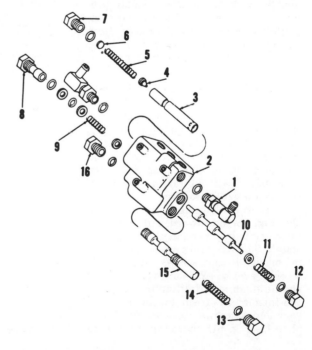

Fig. 59—Exploded view of solenoid and pilot valve assembly, interchangeable in all three locations.

1. Nut
2. Solenoid
3. Armature
4. "O" ring
5. Spring
6. Plunger
7. Back-up ring
8. "O" ring
9. Spring pin
10. Bushing and valve
11. "O" ring
12. Backup ring

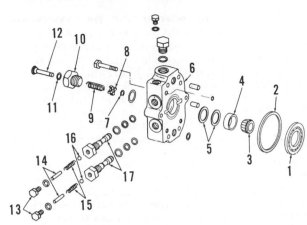

Fig. 61—Exploded view of Power Front Drive Wheel manifold showing component parts.

1. Valve plate
2. Packing
3. Bearing cone
4. Bearing cup
5. Shims (0.003 & 0.010)
6. Manifold
7. Snap ring
8. Retainer
9. Spring
10. Valve plug
11. "O" ring
12. Valve
13. Plug
14. Spring pin
15. Spring
16. Check valve ball
17. Valve body

block (5—Fig. 62) then use suitable pullers to remove inner bearing cone (3 —Fig. 61). Motor shaft, outer bearing cone and swash plate can now be removed. (see preceding NOTE.)

Before disassembling motor, note the splines on ball guide and cylinder block. If splines do not have a master spline, mark both surfaces for reassembly.

To disassemble the motor, grasp outer diameter of slipper retainer (2— Fig. 62) and remove retainer and nine pistons (1). Remove ball guide (3) and withdraw nine slipper retainer springs (4). See Fig. 63 for method of removal and stacking of pistons for maximum cleanliness.

Wheel motor spring (8—Fig. 62) must be slightly compressed to remove or install retaining ring (11). Either of two methods is acceptable:

(1) Select a 3/8-inch bolt long enough to extend through cylinder block, nut and large flat washers. Install bolt

through cylinder block as shown in Fig. 64 (washer on each end) and compress the spring by tightening nut, then unseat and remove retaining ring (11 Fig. 62).

NOTE: Retaining ring should be pulled (not pried) out of its groove. Prying will probably cause a burr to be turned up on the lapped surface, resulting in oil leakage.

(2). As an alternate method, place cylinder block in a press on wood blocks, retaining ring up. Use a step

plate on spring retainer and slightly compress the spring to remove retaining ring.

When using either method, release pressure slowly after removing retaining ring using extreme care to keep from scratching or scoring polished, machined face of block.

Clean all parts and inspect for excessive wear or other damage. Check pistons and bores in cylinder block for

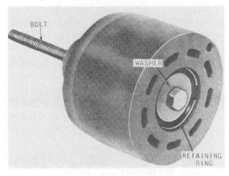

Fig. 64—Use a 3/8-inch, full threaded bolt, flat washers and nut to disassemble motor spring as shown.

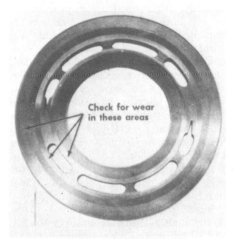

Fig. 65—Inspect steel valve plates for wear or scoring in areas shown.

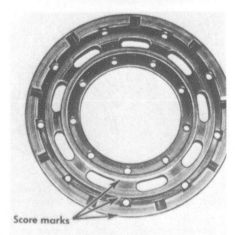

Fig. 66—Inspect brass bearing plates as indicated.

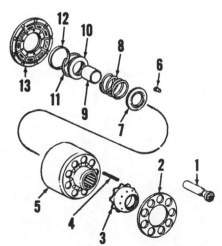

Fig. 62—Exploded view of front wheel drive motor.

1. Piston
2. Slipper retainer
3. Ball guide
4. Springs
5. Cylinder block
6. Aligning dowel
7. Spring seat
8. Spring
9. Spring guide
10. Spring retainer
11. Retaining ring
12. Centering ring
13. Brass bearing plate

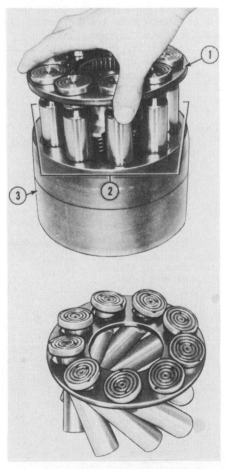

Fig. 63—Method of removal and convenient stacking of pistons.

1. Retainer   2. Pistons   3. Cylinder block

linear scratches and excessive wear. Check cylinder block face for shiny streaks, indicating high pressure leakage between cylinder block and brass bearing plate. Inspect piston slippers for scratches, imbedded material or other damage. Light scratches in slippers can be removed by lapping. All nine slippers must be within 0.002 inch of each other in thickness.

Lubricate all parts with hydraulic fluid and reassemble by reversing the disassembly procedure. Spring seat (7—Fig. 62) must be installed with beveled side away from cylinder block spring (8) and spring retainer (10) installed with retaining groove away from spring. When installing ball guide (3), align master spline of ball guide and cylinder block. On some motors, a prick punch mark is used on one outer tang of ball guide which aligns with a corresponding mark on cylinder block face.

Inspect the steel valve plate (1—Fig. 61) and brass bearing plate (13—Fig. 62) for excessive wear or scoring in areas shown in Figs. 65 and 66. Although both plates are available individually, it is good shop practice to renew both plates whenever one is damaged, to assure proper sealing necessary for efficient operation.

Before installing motor assembly and shaft in motor housing (4—Fig. 68), check wheel hub bearings (10 and 11) for proper adjustment. Shims (13) behind outer bearing cup are available in 0.003 inch and 0.010 inch thickness to allow for an adjustment range of from 0.004 inch preload to 0.002 inch end play. Preload may be considered correct when a pull of 3 pounds at wheel hub is required to turn the hub.

Install outer bearing cone (8) on motor shaft (7). Install motor assembly and swash plate (3) on motor shaft and press inner bearing cone (3—Fig. 61) onto motor shaft. Thoroughly lubricate motor and shaft assembly and carefully install into motor housing (4—Fig. 68), with swash plate (3) installed thick side toward rear of tractor and brass bearing plate (13—Fig. 62) protecting lapped surface of cylinder block. Refer to Fig. 61. Lubricate valve plate (1) and install before oil manifold is bolted to housing. Spring (8—Fig. 62) will resist as the manifold is installed, so tighten evenly to avoid binding. Tighten manifold cap screws to a torque of 35 ft.-lbs. Motor shaft end play should be 0.006-0.012 inch and can be measured with a dial indicator by removing relief valve plug (10—Fig. 61) and associated parts. Shims (5) are available in thicknesses of 0.003-0.010 inch and are used for adjustment if necessary. Remove bearing cup (4) and adjust by adding or removing shims as required.

40. PRIMARY PLANETARY UNIT. The primary planetary unit and secondary sun gear are shown exploded in Fig. 67. Primary ring gear is carried in planetary housing (29—Fig. 68).

Each planet pinion (7—Fig. 67) rides on seventeen (17) loose needle rollers (6). Secondary sun gear (1) splines into hub of planet carrier (8), and serves as pilot for primary sun gear (2). Primary planetary unit may be withdrawn after removing retaining cover (10). Tighten cover retaining cap screws to a torque of 35 ft.-lbs. when installing.

41. SECONDARY PLANETARY AND DRIVE BRAKE UNITS. Refer to Fig. 68 for an exploded view of wheel drive unit. Secondary planet gears (28) are carried in planetary housing (29) which also contains primary ring gear. Secondary ring gear (22) floats in housing until hydraulic pressure is applied by the nine brake pistons (18).

Each planet pinion (28) rides on twenty-two loose needle rollers on shaft (27). Shaft (27) is keyed to housing (29) and carrier plate (24) by steel balls (26). Tighten cap screws retaining carrier plate (24) to a torque of 35 ft.-lbs. when unit is reassembled.

Brake backing plate (23) attaches to planetary piston housing (15) with 18 cap screws and encloses planetary ring gear (22), brake facings (21), brake pressure plate (19) and six separator springs (20). Backing plate must be removed for renewal of brake parts, including the nine brake pistons (18).

NOTE: One of the nine brake pistons should contain a bleed valve, and should be installed slotted end out in the second hole counter-clockwise from the word "UP", which is cast into the brake piston housing.

Tighten brake backing plate attaching screws to a torque of 35 ft.-lbs. when unit is assembled. Mount a dial indicator on outboard side of brake backing plate (23) and check free play of planetary ring gear (22) at three places around internal teeth (inside) circumference. Free play should be at least 0.014 inch.

Tighten planetary brake piston housing retaining cap screws to a torque of 45 ft.-lbs. when unit is reassembled.

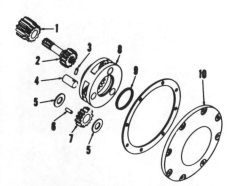

Fig. 67—Exploded view of front wheel drive cover, primary planet carrier and associated parts.

1. Secondary sun gear
2. Primary sun gear
3. Dowel
4. Pinion shaft
5. Thrust washer
6. Bearing roller
7. Primary pinion
8. Planet carrier
9. Snap ring
10. Cover

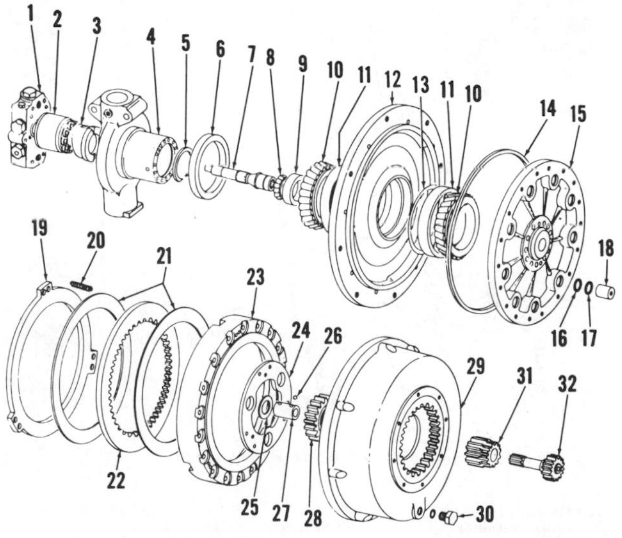

Fig. 68—Exploded view of wheel drive unit showing component parts.

1. Wheel manifold
2. Motor
3. Swashplate
4. Motor housing
5. Snap ring
6. Oil seal
7. Motor shaft
8. Bearing cone
9. Bearing cup
10. Bearing cone
11. Bearing cup
12. Wheel hub
13. Shim
14. Packing
15. Piston housing
16. Quad ring
17. Back-up ring
18. Brake piston
19. Pressure plate
20. Separator spring
21. Brake facings
22. Planetary ring gear
23. Backing plate
24. Retainer
25. Thrust washer
26. Locking ball
27. Pinion shaft
28. Planet pinion
29. Planetary housing
30. Drain plug
31. Secondary sun gear
32. Primary sun gear

ENGINE

All models are equipped with a six cylinder engine; however, many differences will be noted.

4030— 303 CID Gasoline or
        329 CID Diesel
4230— 362 CID Gasoline or
        404 CID Diesel
4430— 404 CID Turbocharged Diesel
4630— 404 CID Turbocharged/
        intercooled Diesel

REMOVE AND REINSTALL

All Models

45. To remove the engine and clutch assembly as a unit, first drain cooling system and, if engine is to be disassembled, drain oil pan. Remove vertical air stack, muffler, hood, cowl, grille screens and engine side panels, Discharge brake accumulators by opening the right brake bleed screw and depressing right brake pedal for about 60 seconds.

Make a clutch split by disconnecting batteries and cables, tachometer cable, heater hoses, throttle linkage, electric wiring connectors on right cowl and inside control support housing and hydraulic lines on right side of engine. Disconnect fitting to cold starting pipes and disconnect wiring and hydraulic lines on left side. On Perma-Clutch models, place a drain pan under housing to catch the oil as engine is separated from housing. Remove the 3-point hitch center link attaching bracket at rear of transmission housing, remove the large plug behind bracket, and pull the hex shaft that extends all the way to the crankshaft and drives the transmission pump. Shaft can become bent if not removed. If equipped with Power Front Wheel Drive, disconnect drain pipe on left side. If equipped with air conditioning, break connections in hoses at couplers on left side bracket. Hold coupler body with one wrench and remove coupler with another wrench. If escaping gas can be heard, tighten coupler and loosen again. Suitably support both halves of tractor separately, remove cap screws securing engine to clutch housing and roll rear half of tractor away from engine and front unit.

NOTE: If tractor is equipped with Sound-Gard body, it will be necessary to remove floor mat, floor panel and filler panels for access to clutch housing cap screws.

Remove wiring harness on engine if necessary, and disconnect hydraulic pump coupling and support. Disconnect fuel line to fuel pump after closing

shut-off valve and remove fuel return pipe. Remove air intake pipe and disconnect both radiator hoses. Remove air conditioning compressor and hoses as an assembly if so equipped. Install JDG-1 engine sling and lifting brackets to a hoist, remove side frame cap screws to engine and slide engine out of side frames.

NOTE: When engine is removed, front unit may be heavy in front therefore unstable. Remove front end weights, if used, and securely support front frame to prevent forward tipping.

Reassemble tractor by reversing the disassembly procedure. Tightening torques are as follows:
Hydraulic pump drive . . . . . . .30 ft.-lbs.
Hydraulic pump support . . . . .85 ft.-lbs.
Side frame to cyl. block . . . . .275 ft.-lbs.
Cylinder block to clutch
    housing . . . . .300 ft.-lbs.

CYLINDER HEAD

All 4030 Models

46. To remove the cylinder head, drain cooling system and remove side panels, grille screens, muffler and hood. Remove air intake pipes up to the intake manifold or carburetor. Remove thermostat by-pass hose, upper radiator hose and the water manifold. Disconnect interfering linkage, then unbolt and remove the intake and exhaust manifolds. On diesel models, remove injector lines and injectors. On all models, remove vent tube, rocker arm cover, rocker arm assembly, then unbolt and remove the cylinder head.

NOTE: Make sure that cylinder liners are held down with bolts and washers if engine crankshaft is to be turned.

Install cylinder head gasket dry. Coat threads of cylinder head retaining cap screws with oil before installation and be sure that hardened flat washers are installed under head of all cylinder head retaining screws. Tighten the cylinder head retaining screws evenly to a final torque of 110 ft.-lbs. torque using the sequence shown in Fig. 69. Tighten the rocker arm clamp screws to 35 ft.-lbs. torque. Retorque cylinder head and readjust valve clearance after engine has run at 1900-2100 rpm for ½-hour. Use same sequence as shown in Fig. 69 and tighten screws to 110 ft.-lbs. torque. Refer to paragraph 51 for adjusting valve clearance.

All 4230, 4430 and 4630 Models

47. To remove the cylinder head, first drain cooling system and remove side panels, grille screens, air stack, muffler and hood. Remove air intake pipes up to turbocharger (4430 and 4630), or to intake manifold or carburetor (4230). Remove thermostat bypass tube, water manifold and upper radiator hose. Remove injector fuel lines and injectors from diesel engines. On all models, disconnect interfering linkage and lines and remove intake and exhaust manifolds. Remove crankcase vent tube, rocker arm cover and rocker arm assembly. Unbolt and remove cylinder head.

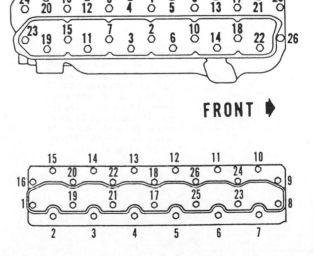

Fig. 69—Use the sequence shown to tighten cylinder head retaining screws. The drawing at top is for 4030 gasoline and diesel models, lower drawing is for all other models.

NOTE: Make sure that cylinder liners are held down with bolts and washers if engine crankshaft is to be turned.

The cylinder head gasket is available in two different thicknesses for diesel models. The red gasket is 0.044 inch thick and the gray gasket is 0.050 inch thick. The correct gasket for use will depend upon piston height compared to head gasket surface of cylinder block. To measure, use a dial indicator with a magnetic base. Be sure that gasket surface of block is clean and flat, then zero the dial indicator with stem against surface of block. Reposition magnetic base so that stem is against top of piston and measure piston height. If the measured height of all pistons is less than 0.009 inch above surface of block, use the red gasket, If any piston is 0.009 to 0.0014 inch above surface of block, use a gray cylinder head gasket. No piston should stand out more than 0.014 inch above block surface. If measurement indicates excessive stand out, determine cause before completing assembly.

On both gasoline and diesel models, install cylinder head gasket dry, but oil threads of cylinder head retaining cap screws. Be sure to install a flat hardened washer on all cap screws.

Refer to lower diagram in Fig. 69 and tighten center cap screw (No. 17) to 105 ft.-lbs. first to hold cylinder head in correct position during tightening sequence. Starting at No. 1 cap screw, tighten remainder of screws in sequence shown in Fig. 69 to 105 ft.-lbs. Retighten all screws in same sequence to 115 ft.-lbs. Tighten rocker arm clamp screws to 55 ft.-lbs.

After operating engine for approximately 45 minutes, retorque cylinder head cap screws (engine hot) to 130 ft.-lbs. using same sequence shown in Fig. 69. Readjust valve clearance as outlined in paragraph 51 and complete assembly.

VALVES AND SEATS

All Models

50. The valve face angle is 44½ degrees on both diesel and gasoline 4030 and 4230 models and the valve seat angle is 45 degrees. Valve face and seat angles are 30 degrees for turbocharged 4430 and 4630 models.

Some models are originally equipped with hardened steel inserts which are pressed into machined bores in cylinder head. On 4030 diesel and gasoline models, the intake valve should be recessed 0.023-0.047 inch and exhaust should be recessed 0.038-0.0720 inch below flush with gasket surface of cylinder head. On 4230 diesel models, the intake valve should be recessed 0.036-0.055 inch, and the exhaust valve should be recessed 0.054-0.068 inch. Turbocharged 4430 and 4630 models should have both intake and exhaust valves protrude beyond gasket surface of cylinder head 0.024-0.038 inch. The manufacturer recommends installation of new valve and seat insert if valve is recessed more than 0.006 inch on turbocharged models.

Intake and exhaust valve stem diameter is 0.3715-0.3725 inch and recommended stem to guide clearance is 0.0020-0.0040 inch. Guides can be knurled to provide correct clearance if clearance is less than 0.010 inch. If clearance exceeds 0.010 inch, bores in guides should be reamed to correct oversize and new valve with larger (oversize) stem fitted. Be sure to reseat valve after guide has been knurled or reamed oversize. Valve seat width should be 0.0781-0.0937 (5/64-3/32) inch for 4030 Diesel models; 0.0625-0.0781 (1/16-5/64) inch for 4030 gasoline models; 0.0830-0.0930 inch for all other models.

Valve clearance (tappet gap) should be adjusted using procedure outlined in paragraph 51.

51. TAPPET GAP ADJUSTMENT. The two-position method of valve tappet gap adjustment is recommended. Refer to Fig. 70 for 4030 models, and Fig. 71 for all other models and proceed as follows:

Turn engine crankshaft by hand using JDE-81 or similar engine rotation tool until number 1 and 6 cylinders are at TDC. This TDC location is determined by using timing pin in flywheel housing to engage hole in flywheel on Diesel models; or by aligning the timing marks on front damper and timing gear housing of gasoline models. Check the valves to determine whether front or rear cylinder is at top of compression stroke. (Exhaust valve on adjacent cylinder will be partly open). Use the appropriate diagram (Fig. 70 or 71) and adjust the indicated valves; then turn crankshaft one complete turn until "TDC" timing mark is again aligned. Adjust remainder of valves using the other diagram. Recommended valve tappet gap is as follows:

4030
Intake (all) . . . . .0.014 in.
Exhaust (Gasoline) . . . . .0.022 in.
  (Diesel) . . . . .0.018 in.
4230 Gasoline
Intake .0.015 in.
Exhaust. . . . . .0.028 in.

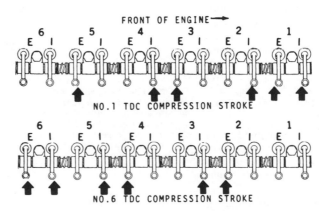

Fig. 70—Adjust the valves on 4030 models as indicated by arrows in upper half of Fig. when No. 1 piston is at TDC on compression stroke. Tappet gap should be as listed in table in text. Refer to lower half of Fig. for remainder of valves.

All Diesel Models Except 4030
Intake .0.018 in.
Exhaust . . . . .0.028 in.

VALVE ROTATORS

All Models

52. Positive type valve rotators are originally installed only on exhaust valves of 4030 gasoline, 4230 gasoline and early 4230 diesel models. 4030 Diesel models were not originally equipped with rotators. All other models were originally equipped with valve rotators on all (intake and exhaust) valves.

Normal service consists of renewing the complete unit. Rotators can be considered satisfactory if the valve turns a slight amount each time it opens.

VALVE GUIDES AND SPRINGS

All Models

53. Valve guide bores are an integral part of cylinder head. Standard valve guide bore diameter is 0.3745-0.3755 inch and normal stem clearance is 0.002-0.004 inch. If stem to guide clearance is 0.006-0.010 inch, the guide can be knurled; however, if clearance exceeds 0.010 inch, resize guide and install valve with correct oversize stem.

Intake and exhaust valve springs are interchangeable and may be installed either end up. Renew any spring which is distorted, rusted or discolored, or does not meet the test specifications which follow:

Free Length (approx.) . . . . .2.12 in.
Lbs. Test at Length (Inches)
   Closed . . . . .54-62 at 1.81 in.
   Open . . . . .133-153 at 1.36 in.

ROCKER ARMS

All Models

54. The rocker arm shaft attaches to bosses which are cast into cylinder head and is held is place by clamps. Shaft rotation is prevented by a spring pin in cylinder head which enters a hole in shaft for positive positioning of lubrication passages.

Rocker arms are right hand and left hand assemblies, except for 4030 gasoline and diesel models which use 12 identical rocker arms. Recommended clearance of rocker arms to shaft is 0.0005-0.0035 inch. Bushings are not available; if clearance is excessive, renew rocker arms and/or shaft.

When reassembling, make sure spring pin aligns with locating hole in shaft, tighten clamp screws to correct torque (35 ft.-lbs. for 4030 models; 55 ft.-lbs. for all other models), then adjust tappet gap as outlined in paragraph 51.

CAM FOLLOWERS (TAPPETS)

All Models

55. The mushroom type cam followers can be removed from below after removing camshaft as outlined in paragraph 58. The cam followers operate in unbushed bores in engine block and are available in standard size only.

TIMING GEAR COVER AND CRANKSHAFT FRONT OIL SEAL

All Models

56. To remove the timing gear cover, first drain cooling system and remove hood, grille screens and engine side panels. Remove radiator and fan shroud from left side after disconnecting oil cooler from radiator.

Remove pressure and return lines from hydraulic pump, disconnect pump drive shaft and coupler and remove pump and support. Loosen fan belts, remove crankshaft damper pulley using a suitable puller, remove cover retaining cap screws and lift off the cover. The lip type front oil seal is supplied in a kit which also includes a steel wear sleeve which is pressed on crankshaft in front of gear as shown in Fig. 72. Score old sleeve lightly with a blunt chisel and pry sleeve from shaft. Coat inner surface of new sleeve with a non-hardening sealant and install with a suitable screw-type installer such as JDE-3. Install oil seal in cover from inside. Sealing lip should be toward the rear and seal should bottom in its bore.

When installing timing gear cover, tighten retaining cap screws to a torque of 30 ft.-lbs. and damper pulley retaining cap screw to 150 ft.-lbs. Complete the assembly by reversing the disassembly procedure. Other tightening torques are given in paragraph 45.

TIMING GEARS

All Models

57. On all models except 4030, the timing gear train consists of crankshaft gear, camshaft gear, injection pump drive gear (which fits on camshaft directly behind camshaft gear) and injection pump shaft gear as shown in Fig. 72. Fig. 73 shows 4030 model gear train which also has an upper and lower idler gear, and oil pump drive gear. Gears are available in standard size only. If backlash is excessive, renew the parts concerned. On 4030 mod-

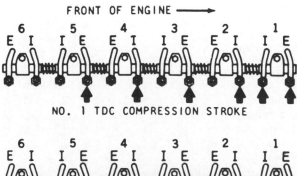

Fig. 71—On all models except 4030, adjust the valves indicated by arrows in upper half of Fig. when No. 1 piston is at TDC on compression stroke. Tappet gap should be as listed in table in text. Refer to lower half of Fig. for remainder of valves.

Fig. 72—Score oil seal wear sleeve as shown using a blunt chisel, then pry from shaft when renewal is indicated. Timing marks align as shown.

els the camshaft must be timed with the mark on rim of gear closest to the crankshaft gear, and on a line which passes through the center of camshaft and crankshaft. Use timing tool JD254 to establish accurate centerline. On all other models "V" marks on camshaft gear must align with crankshaft gear (Fig. 72). On all models, Number 1 cylinder must be at TDC when aligning timing marks. When installing crankshaft gear, heat gear to approximately 350 degrees F. using a hot plate or oven and install with a press or JDH-7 driver with timing mark to front.

CAMSHAFT AND BEARINGS

All Models

58. To remove the camshaft, first split tractor at front of engine as follows:

Drain cooling system, remove air stack and muffler, grille screens, hood, left tractor step, batteries and boxes. Remove air conditioning compressor if so equipped, and disconnect front couplers by using two wrenches on fittings. If gas escapes, tighten couplers and loosen them again until no gas escapes. Lay compressor and hoses aside as a unit. Disconnect hydraulic pump inlet and outlet pipes, pump coupler and support. Remove air intake pipes to carburetor, intake manifold or turbocharger. Close fuel shut-off and disconnect supply pipe to fuel pump. On diesel models disconnect fuel return line. Disconnect steering pipes, oil cooler return pipe, radiator and heater hoses. If equipped, disconnect Power Front Wheel drain pipe. Separate front wiring harness at connectors. Support front assembly to prevent tipping, install JDG-1 engine sling, and remove side frame to engine cap screws. Roll rear section away from front end and

place a support under clutch housing.

Remove timing gear cover. Remove oil pump as in paragraph 70 on all models except 4030.

Remove rocker arm cover, rocker arms assembly and push rods. Raise and secure cam followers in their uppermost position using magnetic holders or other suitable means. Remove speed-hour meter drive from right side of engine (left rear upper flywheel housing on 4030 models), fuel supply pump, and distributor on gasoline models. Working through openings provided in camshaft gear, remove the four cap screws (two cap screws on 4030) securing camshaft thrust plate to front face of engine block, then withdraw camshaft and gear assembly forward out of engine.

4030 models have camshaft journals which should measure 2.1997-2.2007 inches, and all other models have 2.3745-2.3755 inch camshaft journals. All models should have a clearance of 0.002-0.005 inch in bushings. The presized copper lead camshaft bushings are interchangeable. To install bushings after camshaft is out, detach cylinder block from clutch housing, remove clutch, flywheel and camshaft bore plug, then pull bushings into block bores using a piloted puller. Make sure oil supply holes in block are aligned with holes in bushings. The elongated hole in bushing goes to the top.

Camshaft end play of 0.0025-0.0085 inch is controlled by the thickness of camshaft thrust plate. End play can be measured with a dial indicator when camshaft is installed, or with a feeler gage when camshaft unit is out. Thrust plate thickness is 0.187-0.189 inch on all models except 4030, which is 0.156-0.158 inch.

Camshaft gear and injector pump drive gear if so equipped can be removed with a press when camshaft is out, after removing the retaining cap screw and washer. (Gears for 4030 models simply press off, with no cap screw. Speed-hour meter drive shaft can be removed if necessary by threading exposed end, installing a nut and pulling shaft from camshaft.) Align key slot in gear with Woodruff key in shaft, make sure thrust plate and spacer are installed and press gear on shaft until it bottoms (4030 does not have a spacer). Tighten gear retaining cap screw to a torque of 85 ft.-lbs. and thrust plate retaining cap screws to a torque of 20 ft.-lbs. on all models except 4030 model which torques to 35 ft.-lbs. Install upper idler gear on 4030 with thrust washers and cap screws, and torque to 65 ft.-lbs. Align timing marks as outlined in paragraph 57.

ROD AND PISTON UNITS

All Models

59. Connecting rod and piston units are removed from above after removing cylinder head, oil pan and rod bearing caps. When reinstalling, correlation numbers, small and large slots and tangs on rod and cap must be in register. Rods and head of piston are stamped "FRONT" for proper installation. Use new connecting rod cap screws and dip in oil when installing. Tighten connecting rod cap screws to a torque of 65 ft.-lbs. on 4030 models and 100-110 ft.-lbs. on all other models.

NOTE: Do not rotate crankshaft with head removed nor attempt to remove rod and piston units without first bolting liners down using washers and short cap screws as shown in Fig. 74.

PISTONS, RINGS, PINS AND SLEEVES

All Models

60. **PISTONS AND RINGS.** All models are equipped with aluminum alloy, cam ground pistons which use two compression rings and one oil control ring. All rings are located above the piston pin.

Gasoline engines used in 4030 and 4230 models use conventional piston rings with straight sides which are perpendicular to cylinder bore. Piston ring side clearance in groove should be less than 0.005 inch for 4030 Gasoline models; less than 0.010 inch for 4230 Gasoline models.

Diesel Model 4030 and 4230 pistons have keystone type piston ring in top groove, and conventional rectangular ring in second groove. A special ring groove wear gage (JDE-62) is necessary for checking the top ring groove for wear. A new second compression ring should have less than 0.005 inch

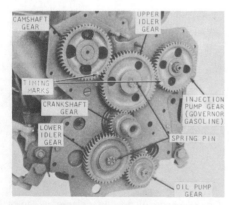

Fig. 73—4030 model timing gears and injection pump gears with cover removed, showing timing marks.

Fig. 74—Lock sleeves in position using a cap screw and washer as shown, when cylinder head is removed for engine service.

side clearance in groove of 4030 Diesel models; less than 0.010 inch side clearance in second groove of 4230 Diesel models. Side clearance of oil ring should be 0.0025-0.0040 inch and should not exceed 0.010 for 4230 Diesel models.

All 4430 and 4630 models have keystone compression rings in the top two grooves of pistons. A special ring groove wear gage (JDE-55) should be used to check for wear. A new oil control ring should have 0.0024-0.0040 inch side clearance in groove and piston should be renewed if side clearance exceeds 0.0065 inch.

Pistons and sleeves are selectively fitted on some models. Pistons marked "L" should only be used in sleeves which are stamped "LL" or "LV" on the sleeve flange. Pistons marked "H" should be installed in sleeves with "HH" or "HV" stamped on flange.

The manufacturer recommends cleaning pistons using Immersion Solvent "D-Part," and Hydra-Jet Rinse Gun or Glass Bead Blasting Machine.

61. SLEEVES, PACKING AND "O" RINGS. The renewable wet type cylinder sleeves are available in standard size only; however, some sleeves and pistons are selectively fitted. Pistons marked "L" should only be used in sleeves with "LL" or "LV" stamped on flange. Pistons marked "H" should only be used in sleeves stamped "HH" or "HV".

Sleeve flange at upper edge is sealed by the cylinder head gasket. Sleeves are sealed at lower edge by packing shown in Fig. 75. Sleeves normally require loosening using a sleeve puller, after which they can be withdrawn by hand. Out-of-round or taper should not exceed 0.005 inch. If sleeve is to be reused, it should be deglazed using a normal cross-hatch pattern.

When reinstalling sleeves, first make sure sleeve and block bore are absolutely clean and dry. Carefully remove all rust and scale from seating surfaces and packing grooves in block and from areas of water jacket where loose scale might interfere with sleeve or packing installation. If sleeves are being reused, buff rust and scale from outside of sleeve.

Install sleeve without the seals and measure standout. Check sleeve standout at several locations around sleeve. Also check to be sure that sleeve will slip fully into bore without force. If sleeve cannot be pushed down by hand, recheck for scale or burrs. Sleeve stand-out should be 0.000-0.004 inch for used cylinder block, but can be 0.002-0.005 inch for new cylinder block.

Check fit of other sleeves, if standout is less than specified. If stand-out is excessive, check for scale or burrs, then, if necessary, select another sleeve. After matching sleeves to all the bores, mark the sleeves then refer to the appropriate following paragraphs for packing and sleeve installation.

Apply liquid (lubricating) soap such as part number AR54749 to the square section ring (1—Fig. 75) and install over lower end of cylinder liner (sleeve). Slide the square section ring up against shoulder on sleeve, make sure that ring is not twisted and that longer slides are parallel with side of sleeve as shown. Apply the liquid soap to round section "O" rings (2) and install in grooves in cylinder block. If some of the "O" rings are red and some are black, the red "O" rings should be installed in top groove in block and black "O" rings should be in lower groove. Be sure that "O" rings are completely seated in grooves so that installing the sleeve will not damage the "O" rings. Observe the previously affixed mark indicating correct cylinder location, then install sleeves carefully into correct cylinder block bore. Also check the sleeve flange for marks "LL" or "HH". If flange is stamped, install sleeve with stamped "LL" or "HH" mark toward front of engine. Work sleeve gently into position by hand until it is finally necessary to tap sleeve into position using a hardwood block and hammer.

NOTE: Be careful not to damage the packing rings. Check the sleeve standout (Fig. 75) with packing installed. The

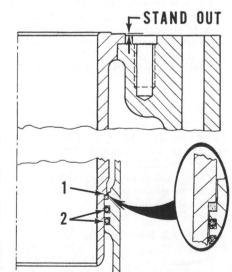

Fig. 75—Cross-section of cylinder sleeve showing square section packing (1) and round "O" rings (2). Refer to text for correct installation of cylinder sleeves.

difference between this measured standout and similar measurement taken earlier for same sleeve in same bore without packing will be the compression of the packing. If compression is less than 0.0126 inch, the square section packing ring will not seal properly. Remove sleeve from cylinder block, check packing to be sure that installation has not cut the packing ring. If shoulders on sleeve and in cylinder block do not provide proper compression of the square section packing ring, install different sleeve and recheck. If a different sleeve will not provide enough compression of packing sleeve and correct packing is installed, suggested repair is to install new cylinder block.

62. SPECIFICATIONS. Specifications of pistons and sleeves are as follows:

4030 Gasoline
Piston Skirt
  Diameter*.........3.8571-3.8581 in.
Cylinder Bore
  Diameter...........3.8581-3.8595 in.
Piston Skirt* to Cylinder
  Clearance ....0.0000-0.0034 in.
Piston Pin Diameter...1.1875-1.1879 in.
Piston Pin Bore in
  Piston ....1.1880-1.1883 in.

4030 Diesel
Piston Skirt
  Diameter*.........4.0125-4.0135 in.
Cylinder Bore
  Diameter...........4.0150-4.0164 in.
Piston Skirt* to Cylinder
  Clearance ....0.0015-0.0039 in.
Piston Pin Diameter—
  Early Models ....1.1871-1.1879 in.
  Later Models ....1.3748-1.3752 in.
Piston Pin Bore in Piston—
  Early Models ....1.1880-1.1883 in.
  Later Models ....1.3753-1.3757 in.

4230 Gasoline
Piston Skirt
  Diameter*.........4.2455-4.2465 in.
Cylinder Bore
  Diameter...........4.2493-4.2507 in.
Piston Skirt* to Cylinder
  Clearance ....0.0028-0.0052 in.
Piston Pin Diameter...1.2497-1.2500 in.
Piston Pin Bore in
  Piston ....1.2503-1.2509 in.

4230 Diesel
Piston Skirt Diameter*—
  Early (Not marked) .4.2455-4.2465 in.
  Marked "L", green .4.2459-4.2466 in.
  Marked "H", yellow .4.2466-4.2473 in.
Cylinder Bore Diameter—
  Early (Not marked) .4.2493-4.2507 in.
  Marked "LV" ......4.2493-4.2500 in.
  Marked "HV"......4.2500-4.2507 in.
Piston Skirt* to Cylinder Clearance—

Early (Pistons and Sleeve not
  marked)..........0.0028-0.0053 in.
Later (Marked "L" & "LV" or "H"
  & "HV") ....0.0027-0.0041 in.
Piston Pin Diameter...1.4997-1.5003 in.
Piston Pin Bore in
  Piston ....1.5003-1.5009 in.

4430 & 4630 Diesel
Piston Skirt Diameter*—
  Marked "L" ....4.2459-4.2466 in.
  Marked "H" ....4.2466-4.2473 in.
Cylinder Bore Diameter—
  Marked "LL" or
    "LV" ....4.2493-4.2500 in.
  Marked "HH" or
    "HV" ....4.2500-4.2507 in.
Piston Skirt* to Cylinder
  Clearance ....0.0027-0.0041 in.
Piston Pin Diameter...1.6247-1.6253 in.
Piston Pin Bore in
  Piston ....1.6253-1.6259 in.

PISTON PINS

All Models

63. The full floating type piston pins
are retained in piston bosses by snap
rings. Pins are often available in over-
sizes as well as standard. Check parts
source for availability.

The recommended fit of piston pins
is a hand push fit in piston bores and a
slip fit in connecting rod bushings.
Standard diameter and clearances are
as follows:

Early 4030 Diesel
Piston Pin Diameter—
  Standard..........1.1871-1.1879 in.
  Wear Limit..............1.1860 in.
Bore in Connecting Rod for Piston
  Pin—
  Standard..........1.1886-1.896 in.
Bore in Piston for Piston Pin—
  Standard..........1.1880-1.1883 in.
  Wear Limit..............1.1893 in.

Late 4030 Diesel
Piston Pin Diameter—
  Standard..........1.3748-1.3752 in.
  Wear Limit..............1.3733 in.
Bore in Connecting Rod for Piston
  Pin—
  Standard..........1.3760-1.3770 in.
Bore in Piston for Piston Pin—
  Standard..........1.3753-1.3757 in.
  Wear Limit..............1.3767 in.

4030 Gasoline
Piston Pin Diameter—
  Standard..........1.1875-1.879 in.
  Wear Limit..............1.1860 in.
Bore in Connecting Rod for Piston
  Pin—
  Standard..........1.1886-1.1896 in.
Bore in Piston for Piston Pin—
  Standard..........1.1880-1.1883 in.

Wear Limit..............1.1893 in.

4230 Diesel
Piston Pin Diameter—
  Standard..........1.4997-1.5003 in.
Bore in Connecting Rod for Piston
  Pin—
  Standard..........1.5010-1.5020 in.
Bore in Piston for Piston Pin—
  Standard..........1.5003-1.5009 in.
Piston Pin to Rod Bushing—
  Desired Clearance...0.0007-0.0023 in.
  Wear Limit..............0.003 in.

4230 Gasoline
Piston Pin Diameter—
  Standard..........1.2497-1.2500 in.
  Wear Limit..............1.2492 in.
Bore in Connecting Rod for Piston
  Pin—
  Standard..........1.2510-1.2520 in.
Bore in Piston for Piston Pin—
  Standard..........1.2503-1.2509 in.
  Wear Limit..............1.2519 in.

4430 Diesel, 4630 Diesel
Piston Pin Diameter—
  Standard..........1.6247-1.6253 in.
Bore in Connecting Rod for Piston
  Pin—
  Standard..........1.6260-1.6270 in.
Bore in Piston for Piston Pin—
  Standard..........1.6253-1.6259 in.
Piston Pin to Rod Bushing—
  Desired Clearance...0.0007-0.0023 in.
  Wear Limit..............0.003 in.

CONNECTING RODS
AND BEARINGS

All Models

64. Connecting rod bearings are
steel-backed inserts. Bearings are available in standard size as well as undersizes of 0.002, 0.010, 0.020 and 0.030
inch. Refer to paragraph 63 for specifications concerning the piston pin bushing.

Mating surfaces of rod and cap have
milled tongues and grooves which posi-
tively locate cap and prevent it from
being reversed during installation. Connecting rods are marked "FRONT" for
proper installation. Check the connecting rods, bearings and crankpin journals for excessive taper and against the
values which follow:

4030 Diesel, 4030 Gasoline
Crankpin Std.
  Diameter..........2.7480-2.7490 in.
Crankpin to Rod Bearing Diametral
  Clearance—
  Desired ....0.0012-0.0042 in.
  Wear Limit..............0.0062 in.
Rod Bolt Torque ....65 ft.-lbs.

4230 Diesel
Crankpin Std.
  Diameter..........2.9982-2.9992 in.
Crankpin to Rod Bearing Diametral
  Clearance—
  Desired ....0.0015-0.0045 in.
  Wear Limit..............0.0065 in.
Rod Bolt Torque........100-110 ft.-lbs.

4230 Gasoline
Crankpin Std.
  Diameter..........2.9985-2.9995 in.
Crankpin to Rod Bearing Diametral
  Clearance—
  Desired ....0.0015-0.0045 in.
  Wear Limit..............0.0065 in.
Rod Bolt Torque.........90-100 ft.-lbs.

4430 Diesel, 4630 Diesel
Crankpin Std.
  Diameter..........2.9980-2.9990 in.
Crankpin to Rod Bearing Diametral
  Clearance—
  Desired ....0.0015-0.0045 in.
Rod Bolt Torque—
  Early-Connecting Rod Marked
    R41423 ....100-110 ft.-lbs.
  Later-Connecting Rod Marked
    R58882 ....110-140 ft.-lbs.

Check crankshaft journals for taper
and out-of-round conditions. Limit for
journal taper is usually considered to
be 0.0001 inch for each one inch of
journal length. Out-of-round limit is
0.0030 inch for 4030 models; 0.0040 inch
for other models.

CRANKSHAFT AND BEARINGS

All Models

65. The crankshaft is supported in
seven main bearings on all engines except 4230 Gasoline model, which has
four main bearings. Crankshaft end
play is controlled by the flanged third
main bearing on 4230 Gasoline models,
by the fifth main bearing on all other
models. Upper and lower bearing shells
may be interchangeable, with both
halves containing oil holes; however, if
only one half of bearing has a hole,
make sure that half is installed in the
block, to insure that oil reaches all
bearings.

All main bearing caps can be removed from below after removing oil
pan and oil pump, and caps are numbered for proper reassembly. When renewing bearings, make sure that locating lug on bearing shell is aligned with
milled slot in cap and block bore. After
caps are loosely installed, bump crankshaft forward and rearward to align
thrust flanges; then tighten main bearing cap screws to a torque of 85 ft.-lbs.
on 4030 models; all other models to 150
ft.-lbs. Main bearings are available in

undersizes of 0.002, 0.010, 0.020 and 0.030 inch. The thrust bearing is available in all undersizes with standard flange width; and in 0.010 inch undersize and 0.007 inch oversize flange width in all models except 4030. Thrust flange thickness should be sufficient to permit crankshaft end play within desired limits.

To remove the crankshaft, first remove engine as outlined in paragraph 45 and proceed as follows: Remove flywheel and crankshaft rear oil seal retainer. Remove crankshaft pulley, timing gear cover, oil pan and oil pump. Remove rod and main bearing caps and lift out crankshaft. The hardened crankshaft front and rear wear sleeves are a press fit on flywheel flange and nose of crankshaft, and are both renewable. When installing, refer to paragraph 56 for front wear sleeve, and paragraph 66 for rear sleeve. Check crankshaft and bearings against the values which are given for crankpin in paragraph 64, and for main bearings as follows:

4030 Diesel, 4030 Gasoline
Crankshaft End Play—
  Desired . . . . 0.0020-0.0080 in.
  Wear Limit . . . . 0.0150 in.
Main Bearing Journal—
  Standard Diameter. . . 3.1230-3.1240 in.
  Desired Clearance in
    Bearing. . . . . . . 0.0016-0.0046 in.
  Wear Limit . . . . 0.0060 in.
  Regrind if Taper
    Exceeds . . . . 0.0010 in./1 in. length
  Regrind if Out-of-Round
    Exceeds . . . . 0.003 in.
Connecting Rod Journal—
  Standard Diameter. . . 2.7480-2.7490 in.
  Desired Clearance in
    Bearing. . . . . . . 0.0012-0.0042 in.
  Wear Limit . . . . 0.0062 in.
Main Cap Torque—
  Engine Serial No. E246640 and
    Earlier. . . . . 85 ft.-lbs.
  Engine Serial No. E246641 and
    Later . . . . 110 ft.-lbs.
Rod Bolt Torque . . . . 65 ft.-lbs.
Crankshaft Damper or Pulley to
  Crankshaft Torque. . . . . . . 85 ft.-lbs.
Flywheel to Crankshaft
  Torque . . . . 85 ft.-lbs.

4230 Diesel, 4230 Gasoline
Crankshaft End Play—
  Desired . . . . 0.0025-0.0085 in.
  Wear Limit . . . . 0.0150 in.
Main Bearing Journal, Serial No.
E395613 and Earlier—
  Standard Diameter. . . 3.3715-3.3725 in.
  Desired Clearance . . . 0.0017-0.0047 in.
  Wear Limit . . . . 0.0077 in.

Regrind if Taper
  Exceeds . . . . 0.0001 in./1 in. length
Regrind if Out-of-Round
  Exceeds . . . . 0.0040 in.
Main Bearing Journal, Serial No.
E395614 and Later—
  Standard Diameter. . . 3.3725-3.3735 in.
  Desired Clearance . . . 0.0007-0.0037 in.
  Wear Limit . . . . 0.0077 in.
  Regrind if Taper
    Exceeds . . . . 0.0001 in./1 in. length
  Regrind if Out-of-Round
    Exceeds . . . . 0.0040 in.
Connecting Rod Journal—
  Standard Diameter,
    Diesel . . . . 2.9982-2.9992 in.
    Gasoline . . . . 2.9985-2.9995 in.
  Desired Clearance in
    Bearing. . . . . . . 0.0015-0.0045 in.
  Wear Limit . . . . 0.0065 in.
Main Cap Torque . . . . 150 ft.-lbs.
Rod Bolt Torque—
  Diesel . . . . 100-110 ft.-lbs.
  Gasoline . . . . 90-100 ft.-lbs.
Crankshaft Damper Torque . . 150 ft.-lbs.
Rear Oil Seal Housing
  Torque . . . . 20 ft.-lbs.
Flywheel to Crankshaft
  Torque . . . . 85 ft.-lbs.

4430 Diesel, 4630 Diesel
Crankshaft End Play—
  Desired . . . . 0.0015-0.0150 in.
  Wear Limit . . . . 0.0150 in.
Main Bearing Journal (4430 before
Serial No. E410333; 4630 before
Serial No. E410265)—
  Standard Diameter . . 3.3715-3.3725 in.
  Desired Clearance . . 0.0017-0.0047 in.
  Wear Limit . . . . 0.0077 in.
  Regrind if Taper
    Exceeds . . . . 0.0001 in./1 in. length
  Regrind if Out-of-Round
    Exceeds . . . . 0.0040 in.
Main Bearing Journal (4430 Serial No.
E41033-E433251; 4630 Serial No.
E410265-E432598)—
  Standard Diameter. . . 3.3725-3.3735 in.
  Desired Clearance . . . 0.0007-0.0037 in.
  Wear Limit . . . . 0.0067 in.
  Regrind if Taper
    Exceeds . . . . 0.0001 in./1 in. length
  Regrind if Out-of-Round
    Exceeds . . . . 0.0004 in.
Main Bearing Journal (4430 Serial No.
after E433251; 4630 Serial No. after
E432598)—
  Standard Diameter. . . 3.3730-3.3740 in.
  Desired Clearance . . . 0.0002-0.0032 in.
  Wear Limit . . . . 0.0062 in.
  Regrind if Taper
    Exceeds . . . . 0.0001 in./1 in. length
  Regrind if Out-of-Round
    Exceeds . . . . 0.0004 in.
Connecting Rod Journal—
  Standard Diameter. . . 2.9980-2.9990 in.
  Desired Clearance . . . 0.0015-0.0045 in.

Main Cap Torque. . . . . 150 ft.-lbs.
Rod Bolt Torque—
  Before 4330 Serial No.
    41033 . . . . 100-110 ft.-lbs.
  Before 4630 Serial No.
    410265 . . . . 100-110 ft.-lbs.
  After 4330 Serial No.
    410332 . . . . 110-140 ft.-lbs.
  After 4630 Serial No.
    410264 . . . . 110-140 ft.-lbs.
Crankshaft Damper . . . . . . 150 ft.-lbs.
Rear Oil Seal
  Housing Torque . . . . 20 ft.-lbs.
Flywheel to Crankshaft
  Torque . . . . 85 ft.-lbs.

CRANKSHAFT REAR OIL SEAL

All Models

66. The crankshaft rear oil seal is contained in a retainer plate which is attached to rear face of cylinder block by cap screws. Seal is available only in a kit which also includes the steel wear sleeve. To renew the seal, first detach (split) engine from clutch housing as outlined in paragraph 166 and remove clutch and flywheel.

Unbolt and remove oil seal retainer plate and score wear sleeve lightly with a dull chisel so as not to damage flywheel flange, and pry wear sleeve from crankshaft using a screwdriver or pry bar. Drive or press the seal from retainer being careful not to damage retainer plate. Install seal with closed side to rear of engine. Install wear sleeve with a suitable driver so that rear edge is flush with rear face of crankshaft flange. Check sealing face runout with a dial indicator. Runout should not exceed 0.006 inch. Tighten screws attaching seal retainer to 20 ft.-lbs torque when correctly centered.

FLYWHEEL

All Models

67. Flywheel is doweled to crankshaft flange and retained by four cap screws. Flywheel can be removed by using forcing screws in tapped holes provided.

To install flywheel ring gear, heat gear evenly to approximately 300 degrees F. and position gear so that chamfered end of gear teeth face toward front of engine. When installing flywheel, align dowel hole in flywheel over dowel in flange, coat threads of retaining cap screws with sealant and tighten evenly to a torque of 85 ft.-lbs.

CLUTCH SHAFT PILOT BUSHING ADAPTER

All Models

68. On 4030 models the clutch shaft pilot bushing is held in an adapter which is pressed into the flywheel. The adapter is also the hex drive for the transmission pump drive shaft. The pilot bushing can be renewed without pressing the adapter from flywheel, and the adapter can be pressed out and renewed if hex drive becomes worn. Bushing should be installed until it is 0.020-0.040 inch below adapter face. If necessary to renew adapter, press new adapter in until flange bottoms in bore of flywheel.

On all models except 4030 with Power-Shift transmission, a bearing puller can be used to pull the sealed pilot bearing from the crankshaft bore after torsional damper is removed.

On all models except 4030 with Perma-Clutch, the clutch shaft pilot bushing is carried in an adapter which is pressed into end of crankshaft. The adapter is also the hex drive for the transmission pump drive shaft. The pilot bushing can be removed using a blind hole puller to withdraw bushing from bore in adapter. If necessary, adapter can be pulled from crankshaft as shown in Fig. 76. Install a bolt and washer which will fit into pilot bushing bore, install a H-214-R snap ring in adapter bore groove to prevent washer from coming out of bore and use type puller shown. Protect the face of new adapter and drive it tight against

crankshaft flange. New bushing should be driven in until it is 0.020-0.040 inch below adapter face.

OIL PAN

All Models

69. Engine oil pan can be removed without interference from other components. Drain engine oil, remove oil filter access plate and filter from side of oil pan. Use a floor jack or other lifting means to remove and install the cast pan. When installing, tighten the 3/8-inch cap screws to a torque of 35 ft.-lbs. and 1/2-inch cap screws to a torque of 85 ft.-lbs.

OIL PUMP

All Models Except 4030

70. **REMOVE AND REINSTALL.** To remove the engine oil pump, first remove oil pan as outlined in paragraph 69, turn crankshaft on gasoline models until No. 1 piston is at TDC on compression stroke; then unbolt and remove oil pump. Correct distributor timing on the gasoline models depends on proper installation of oil pump. When installing the pump on gasoline models only, remove tachometer drive housing and drive gear. If crankshaft was properly timed (TDC-1) when oil pump was removed and crankshaft has not been turned, it may not be necessary to retime distributor. If drive slots will not engage, distributor must be re-

moved or loosened. Refer to Fig. 77. Bottom of tooth which aligns with drive slot should align with a small casting mark just to the side of oil hole in housing as shown.

When pump is installed on gasoline models, "V" mark should be toward crankshaft and drive slot approximately 15° from parallel with crankshaft center line. On diesel models, just make sure that oil pump and camshaft gears mesh properly, and reinstall in reverse order of disassembly. On all models, tighten pump to block attaching screws to 20-25 ft.-lbs. torque. On gasoline models, refer to paragraph 145 and retime ignition distributor.

4030 Models

71. **REMOVE AND REINSTALL.** To remove the engine oil pump from 4030 models, drain and remove oil pan. Remove muffler, air stack, grille screens and hood. Remove oil cooler and necessary pipes. Remove radiator and hoses, hydraulic pump coupler and support. Remove alternator and water pump. Remove crankshaft pulley with suitable puller, and remove timing gear cover. Loosen oil pump drive gear nut and back it off until flush with end of shaft, then tap end of shaft while pulling on gear to free gear from shaft. Remove two cap screws from rear of pump and remove pump and pressure (outlet) pipe. To reinstall, reverse disassembly procedure, tighten drive gear nut to 35 ft.-lbs. torque and stake to shaft. On gasoline models, refer to paragraph 145 and retime ignition distributor.

All Models

72. **OVERHAUL.** To overhaul the removed oil pump, first remove pump cover, remove idler gear and examine pump gears and cover for wear or scoring. Gears are available as a matched set only. Check for wear or excessive looseness between shaft and housing at drive gear end.

Driven gear (Fig. 78) on all models except 4030, is retained to drive shaft by a press fit on shaft. Use a press and suitable mandrel to press shaft out of gear. On 4030 models, remove drive shaft and drive gear as a unit, and

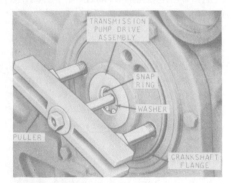

Fig. 76—Method of removing clutch shaft pilot bushing adapter.

Fig. 77—On gasoline models, the oil pump must be properly timed when installing. Refer to text.

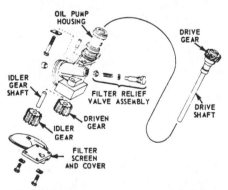

Fig. 78—Exploded view of engine oil pump used on 4230 gasoline model. Other models except 4030 are similar, but do not contain relief valve assembly and filter boss.

Fig. 79—Engine oil pump used on 4030 models.

1. Pickup tube
2. Outlet tube
3. "O" ring
4. Cover
5. Groove pin
6. Gears
7. Drive shaft
8. Idler shaft
9. Pump body
10. Drive gear

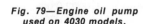

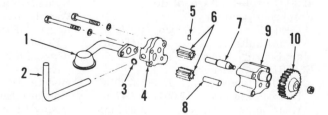

renew shaft, gears and groove pin if necessary. Make sure that groove pin (5—Fig. 79) has not sheared. On all other models, press drive gear on shaft until gear is flush with mating surface of housing. When reassembling, drive gears and idler gears must spin freely and cover must not bind on gears. Re-install pump as outlined in paragraph 70 or 71.

PRESSURE REGULATOR VALVE

4030 Models

73. The pressure regulator valve on 4030 models is located behind the large plug to the right of crankshaft pulley (Fig. 82), in timing gear cover. Correct oil pressure is 50-60 psi with engine at normal operating temperature and 2500 rpm. Shims (Fig. 82) are used to adjust oil relief pressure. Oil filter by-pass valve is contained in the filter itself, operates on a pressure differential of 12-15 psi, and opens in case filter becomes plugged to assure continuing lubrication. Oil cooler by-pass valve is located in oil filter adapter housing, operates on a pressure differential of 15 psi and opens for cold starting or in case of cooler restriction.

All Models Except 4030

74. The oil pressure regulator valve (Fig. 81) is located in filter housing as is the filter by-pass valve (except 4230 gasoline model).

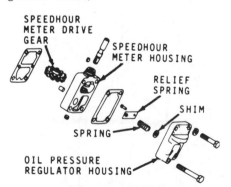

SPEEDHOUR METER DRIVE GEAR
SPEEDHOUR METER HOUSING
RELIEF SPRING
SHIM
SPRING
OIL PRESSURE REGULATOR HOUSING

Fig. 80—Exploded view of speed-hour meter drive housing and associated parts on 4230 gasoline model. Other models do not contain regulator for engine oil.

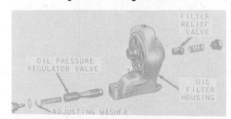

FILTER RELIEF VALVE
OIL PRESSURE REGULATOR VALVE
OIL FILTER HOUSING
ADJUSTING WASHER

Fig. 81—Exploded view of filter housing and associated parts on all models except 4030, 4230 gasoline model, showing relief and regulating valves. Refer to Fig. 80 for 4230 gasoline model, and Fig. 82 for 4030 models.

The 4230 gasoline model regulator valve (Fig. 80) is housed in the speed-hour meter cover, and filter relief valve is part of engine oil pump. Refer to Fig. 78 for exploded view of oil pump, and Fig. 80 for speed-hour meter cover. Refer to Fig. 81 for exploded view of typical filter housing.

Oil from the pump is circulated through the oil cooler (or is by-passed if cooler is restricted), then circulated through filter (or is by-passed if filter is restricted), and finally the oil flows past the oil pressure relief valve which limits the maximum oil pressure. Correct regulated pressure is 40-50 psi when engine is at normal operating temperature and at 1900 rpm. Pressure can be adjusted by adding or removing shims (washers). Adding or removing one shim will change regulated pres-

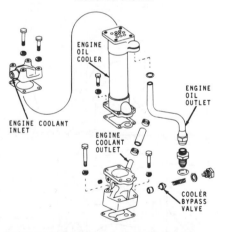

ALUMINUM WASHER
VALVE HEAD
VALVE PLUG
SPRING
SMALL SHIM

Fig. 82—Adjust engine oil pressure on 4030 model as shown.

ENGINE OIL COOLER
ENGINE COOLANT INLET
ENGINE COOLANT OUTLET
ENGINE OIL OUTLET
COOLER BYPASS VALVE

Fig. 83—Exploded view of 4230 gasoline engine oil cooler and associated parts.

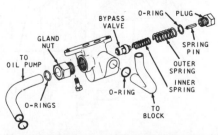

BYPASS VALVE
O-RING
PLUG
GLAND NUT
TO OIL PUMP
SPRING PIN
OUTER SPRING
INNER SPRING
O-RING
O-RINGS
TO BLOCK

Fig. 84—Engine oil cooler by-pass valve used on all diesel models except 4030.

sure by about 5 psi. Oil filter by-pass valve (relief valve) operates on a pressure differential of approximately 15 psi. Filter by-pass valve opens for cold starting, or if filter becomes plugged, to assure continuing lubrication. Oil cooler by-pass valve (Fig. 83 or Fig. 84) operates on a pressure differential of 15 psi and opens for cold starting, or in case of cooler restriction.

OIL COOLER

All Models

75. All models are equipped with an engine oil cooler of the type shown in Fig. 83, 85 or 86. Some variation from

ENGINE OIL COOLER
OIL LINES
COOLANT HOSES
PRESSURE REGULATING VALVE PLUG

Fig. 85—Engine oil cooler used on 4030 models. Cooler by-pass is housed in filter base.

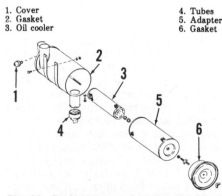

Fig. 86—Exploded view of engine oil cooler typical of type used on all diesel models except 4030. Item 3 is larger on 4630 model, and has a spacer to allow for extra depth.

1. Cover
2. Gasket
3. Oil cooler
4. Tubes
5. Adapter
6. Gasket

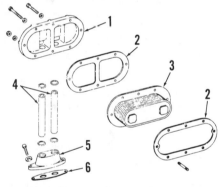

Fig. 87—Exploded view of engine air cleaner used on 4030 and 4230 models. Other models are similar.

1. Warning switch
2. Housing
3. Safety element
4. Unloader valve
5. Primary element
6. Cover

oil coolers shown may be noted on some models. The oil is cooled by the engine coolant liquid which circulates through tubes in cooler body. Oil cooler also serves as attaching point for oil supply line leading to the turbocharger if so equipped. Disassembly for cleaning or other service is normally not required except in cases of contamination of cooling or lubrication systems.

AIR INTAKE SYSTEM

All Models

76. Air used to operate the engine enters through an air stack to the air cleaner. The air then passes through a dual element dry type filter into the carburetor, manifold or turbocharger; then on into the engine. A vacuum switch (1—Fig. 87) connects to a warning indicator lamp which lights to warn the operator if the filters are restricted.

Because of the balanced system and the large volume of air demanded (especially by a turbocharged diesel engine), it is of utmost importance that only the approved parts which are in good condition be used.

To check for air cleaner restriction, install a "T" fitting, an elbow, and the vacuum switch where switch was originally installed, and connect a JDST-11 water vacuum gage in "T" fitting.

Normal vacuum at port for switch (1) with clean filters, air stack extension installed and engine operating at full load is as follows:

Model	Inches of water
4030	5.5 in.
4230	7 in.
4430	10.5 in.
4630	12 in.

Rpm used for test	
4030	2500
All Others	2200

Warning switch closes at a vacuum of 24-26 inches of water (approximately 1.8 inches Hg) and vacuum should never be permitted to be higher than 25 inches of water on any model.

On turbocharged models, intake manifold pressure, checked at the 3/8-inch pipe plug near the inlet tube on intake manifold at 2200 rpm, full load, should be 30-34 inches of mercury (15-17 psi) for 4430; 32-40 inches of mercury (16-20 psi) for 4630 models. A low manifold air pressure may indicate intake manifold air leaks, restricted air intake or air cleaner, exhaust leaks or a malfunctioning turbocharger. Refer to paragraph 84 for service to intercooler

located in intake manifold of 4630 model.

TURBOCHARGER

OPERATION

4430-4630 Models

77. The exhaust driven turbocharger supplies air to the intake manifold at above normal atmospheric pressure. The additional air entering the combustion chamber permits an increase in the amount of fuel burned, and increased power output over an engine of comparable size not so equipped.

The use of the engine exhaust to power the compressor increases engine flexibility, enabling it to perform with the economy of a smaller engine on light loads yet permitting a substantial horsepower increase at full load. Horsepower loss because of altitude or atmospheric pressure changes is also largely reduced.

Because a turbocharger compresses the incoming air, the heat of compression causes the air to expand and become less dense than it would be at a lower temperature. Model 4630 is equipped with an intercooler which lowers the temperature of the intake air which, in turn, increases the density of the intake air. This permits the use of

more fuel and results in more power. Since a greater power output causes more heat, the turbocharger is driven faster, which produces more manifold pressure, and even greater power.

The turbocharger contains a rotating shaft which carries an exhaust turbine wheel on one end and a centrifugal air compressor on the other. The rotating member is precisely balanced and capable of rotative speeds up to 100,000 rpm. Bearings are of the floating sleeve type and the unit is lubricated and cooled by a flow type and the unit is lubricated and cooled by a flow of engine oil under pressure. Exchange turbocharger units are available, or a qualified technician can overhaul the unit if parts are available.

SERVICE

4430-4630 Models

78. In a naturally aspirated diesel engine (without turbocharger) an approximately equal amount of air enters the cylinders at all loads, and only the amount of fuel is varied to compensate for power requirements. Turbocharging may supply up to 3 times the normal amount of air under full load.

All diesel engines operate with an excess of air under light loads. In a naturally aspirated engine, most of the

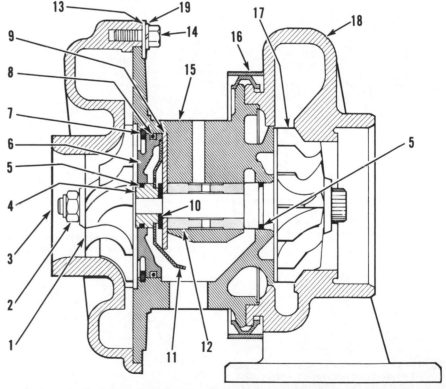

Fig. 90—Cross-sectional view of Schwitzer 3LD turbocharger showing component parts. Refer to Fig. 91 for legend. Other 3LDA and 3LM models are similar.

air is used at full load, and increasing the amount of fuel results in a higher smoke level with little increase in power output. Turbocharging provides a variation of air delivery, and a turbocharged engine operates with an excess of air up to and beyond the design capacity of the engine. When more fuel is provided, the turbocharger speed and air delivery pressure increase, resulting in additional horsepower and heat, with little change in smoke level. Smoke cannot, therefore, be used as a guide to safe maximum fuel setting in a turbocharged engine. **DO NOT** increase horsepower output above that given in **CONDENSED SERVICE DATA** at the front of this manual.

Schwitzer turbocharger models 3LD, 3LDA and 3LM or AiResearch TO-4 units have been used. Be sure to install correct replacement unit if turbocharger is exchanged. The turbocharger consists of the following three main sections. The turbine, bearing housing and compressor.

Engine oil taken directly from the clean oil side of the engine oil filters, is circulated through the bearing housing. This oil lubricates the sleeve type bearings and also acts as a heat barrier between the hot turbine and the compressor. The oil seals used at each end of the shaft are of the piston ring type. When servicing the turbocharger, extreme care must be taken to avoid damaging any of the parts.

CAUTION: DO NOT operate the turbocharger without adequate lubrication. When turbocharger is first installed, or engine has not been run for a month or more, or new oil filter has been installed, turn engine over with starter with fuel cut-off wire disconnected until oil pressure indicator light goes out; reconnect the wire and start engine. Run engine at slow idle speed for at least two minutes before opening throttle or putting engine under load.

Some other precautions to be observed in operating and servicing a turbocharged engine are as follows:

Do not operate at wide-open throttle immediately after starting. Allow engine to idle until turbocharger slows down before stopping engine. This will insure adequate lubrication to the shaft bearings at all times.

Because of increased air flow, care of air cleaner and connections is of added importance. Check the system and condition of restriction indicator whenever tractor is serviced. Make sure exhaust pipe opening is closed and air cleaner connected whenever tractor is transported, to keep turbocharger from turning due to air pressure. If exhaust outlet is equipped with weathercap,

tape the cap closed. If weathercap is missing, use tape to close exhaust outlet pipe.

79. REMOVE AND REINSTALL. To remove the turbocharger, first remove muffler, hood, exhaust elbow and adapter. Remove oil lines and intake hose connections, then unbolt and remove turbocharger from exhaust manifold.

To inspect the removed turbocharger unit, examine turbine wheel and compressor impeller for blade damage, looking through housing and end openings. Using a dial indicator with plunger extension, check radial bearing play through outlet oil port while moving both ends of turbine shaft equally. Check shaft end play with dial indicator working from either end. If end or side play exceeds 0.009 inch; or if any of the blades are broken or damaged, renew or overhaul the unit.

When installing, attach turbocharger to manifold using a new gasket and tighten stud nuts to a torque of 35 ft.-lbs. Install inlet oil line, outlet line, adapter and exhaust elbow after first making sure parts are perfectly aligned. Adapter must have a minimum of 1/16-inch end play and be free to rotate. Undue stress on turbocharger at installation may cause bearing failure. Prime turbocharger as outlined in paragraph 78.

Schwitzer Model 3LD and 3LDA

80. **OVERHAUL.** Remove turbo-

charger as outlined in paragraph 79. Before disassembling, place a row of light punch marks across compressor cover, bearing housing and turbine housing to aid in reassembly. Clamp turbocharger mounting flange (exhaust inlet) in a vise and remove cap screws (14—Fig. 91), lockwashers and clamp plates (13). Remove compressor cover (3). Remove nut from clamp ring (16), expand clamp ring and remove bearing housing assembly (15) from turbine housing (18).

CAUTION: Never allow the weight of the bearing housing assembly to rest on either the turbine or compressor wheel vanes. Lay the bearing housing assembly on a bench so that turbine shaft is horizontal.

Remove locknut (2) and slip compressor wheel (1) from end of shaft. Withdraw turbine wheel and shaft (17) from bearing housing. Place bearing housing on bench with compressor side up. Remove snap ring (7), then using two screwdrivers, lift flinger plate insert (6) from bearing housing. Push spacer sleeve (4) from the insert. Remove oil deflector (11), thrust ring (10), thrust plate (9) and bearing (12). Remove "O" ring (8) from flinger plate insert (6) and remove both seal rings (5) from spacer sleeve and turbine shaft.

Soak all parts in Bendix metal cleaner or equivalent and use a soft brush, plastic blade or compressed air to remove carbon deposits. CAUTION: Do not use wire brush, steel scraper or

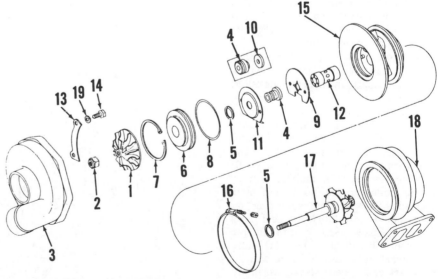

Fig. 91—Exploded view of Schwitzer model 3LD & 3LDA turbocharger assembly used on some models. Parts (4 & 10) shown in inset are for early 3LD models.

1. Compressor wheel	5. Seal rings (not identical)	10. Thrust ring
2. Locknut	6. Flinger plate insert	11. Oil deflector
3. Compressor cover	7. Snap ring	12. Bearing
4. Flinger (Spacer Sleeve)	8. "O" ring	13. Clamp plate
	9. Thrust plate	14. Cap screw
		15. Bearing housing
		16. Clamp ring
		17. Turbine wheel and shaft
		18. Turbine housing
		19. Lockwasher

caustic solution for cleaning, as this will damage turbocharger parts. Glass bead dry blast may be used for cleaning if air pressure does not exceed 40 psi and all traces of the glass beads are rinsed out before assembling.

Inspect turbine wheel and compressor wheel for broken or distorted vanes. **DO NOT** attempt to straighten bent vanes. Check bearing bore in bearing housing, floating bearing (12) and turbine shaft for excessive wear or scoring. Inspect flinger plate insert (6), flinger sleeve (4), oil deflector (11), thrust plate (9) and, if so equipped, thrust ring (10) for excessive wear or other damage. Refer to Fig. 91 and the following for specifications of new parts.

Bearing Bore in Housing
    (15) . . . . 0.8750-0.8755 in.
Bearing (12) Length . . . . . 1.484-1.485 in.
Flinger Sleeve (4)
    Length . . . . 0.517-0.519 in.
    Piston Ring Grooves . . . . 0.064-0.065 in.
Shaft (17) Concentricity—
    Maximum Run-out . . . . 0.001 in.
Shaft (17) Diameter . . . 0.4800-0.4803 in.
Shaft (17) Shoulder
    Length . . . . 1.595-1.596 in.
Thrust Plate (9)
    Thickness . . . . 0.107-0.108 in.
Assembled Bearing Clearances—
    Axial (End Play of
       Shaft . . . . 0.003-0.008 in.
    Radial Clearance Measured at
       Exhaust Turbine
         Blades . . . . 0.018-0.049 in.

Renew all damaged parts and use new "O" ring (8) and seal rings (5) when reassembling. The seal ring used on turbine shaft is copper plated and is larger in diameter than the seal ring used on spacer sleeve. Refer to Figs. 90 and 91 when reassembling.

Install seal ring on turbine shaft, lubricate seal ring and install turbine wheel and shaft in bearing housing. Lubricate I.D. and O.D. of bearing (12), install bearing over end of turbine shaft and into bearing housing. Lubricate both sides of thrust plate (9) and install plate (bronze side out) on the aligning dowels. Install thrust ring (10) and oil deflector (11), making certain holes in deflector are positioned over dowel pins. Install new seal ring on spacer sleeve (4), lubricate seal ring and press spacer sleeve into flinger plate insert (6). Position new "O" ring (8) on insert, lubricate "O" ring and install insert and spacer sleeve assembly in bearing housing, then secure with snap ring (7). Place compressor wheel on turbine shaft, coat threads and back side of nut (2) with graphite grease, or equivalent, then install and tighten nut to 156 in.-lbs. torque. Assemble bearing

housing to turbine housing and align punch marks. Install clamp ring, apply graphite grease to threads, install nut and tighten to 120 in.-lbs. torque. Apply a light coat of graphite grease around machined flange of compressor cover (3). Install compressor cover, align punch marks and secure cover with cap screws, washers and clamp plates. Tighten cap screws evenly to 60 in.-lbs. torque. Fill the oil inlet with engine oil and turn turbine shaft by hand to lubricate bearing and thrust plate.

Check rotating unit for free rotation within the housings. Cover all openings until the turbocharger is reinstalled.

Use a new gasket and install and prime turbocharger as outlined in paragraph 78.

Turbocharger oil supply pressure at 2200 engine rpm should be within 10 psi of engine oil pressure but never less than 25 psi. Minimum return oil flow from turbocharger is ½-gpm at 2200 engine rpm.

Schwitzer Model 3LM

81. **OVERHAUL.** Remove turbocharger as outlined in paragraph 79. Before disassembling, place a row of light punch marks across compressor cover, bearing housing and turbine housing to aid in reassembly. Clamp turbocharger mounting flange (exhaust inlet) in a vise and remove cap screws (14—Fig. 92), lockwashers and clamp plates (13). Remove compressor cover (3). Remove screws (24) then separate

bearing housing assembly (15) from turbine housing (18).

CAUTION: Never allow the weight of the bearing housing assembly to rest on either the turbine or compressor wheel vanes. Lay the bearing housing assembly on a bench so that turbine shaft is horizontal.

Remove locknut (2) and slip compressor wheel (1) from end of shaft. Withdraw turbine wheel and shaft (17) from bearing housing. Remove back plate (20), then remove and discard all of gasket (21). Place bearing housing on bench with compressor side up. Remove snap ring (7), then using two screwdrivers, lift insert (6) from bearing housing. Push spacer sleeve (4) from the insert. Remove oil deflector (11), thrust plate (9) and bearing (12). Remove "O" ring (8) from flinger plate insert (6) and remove both seal rings (5) from spacer sleeve and turbine shaft.

Soak all parts in Bendix metal cleaner or equivalent and use a soft brush, plastic blade or compressed air to remove carbon deposits. CAUTION: Do not use wire brush, steel scraper or caustic solution for cleaning, as this will damage turbocharged parts. Glass bead dry blast may be used for cleaning if air pressure does not exceed 40 psi and all traces of the glass beads are rinsed out before assembling.

Inspect turbine wheel and compressor wheel for broken or distorted vanes. **DO NOT** attempt to straighten bent

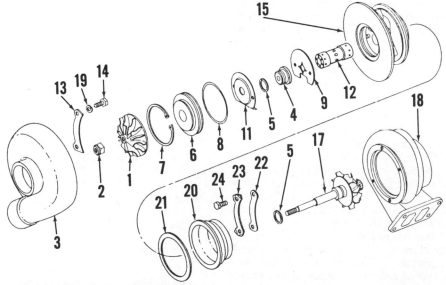

Fig. 92—Exploded view of Schwitzer 3LM turbocharger assembly used on some models.

1. Compressor wheel	6. Insert	13. Clamp plate	19. Lockwasher
2. Locknut	7. Snap ring	14. Cap screw	20. Back plate
3. Compressor cover	8. "O" ring	15. Bearing housing	21. Gasket
4. Thrust sleeve	9. Thrust plate	17. Turbine wheel and	22. Clamp plate
5. Seal rings (not	11. Oil deflector	shaft	23. Lock plate
identical)	12. Bearing	18. Turbine housing	24. Cap screws

vanes. Check bearing bore in bearing housing, floating bearing (12) and turbine shaft for excessive wear or scoring. Inspect flinger plate insert (6), thrust sleeve (4), oil deflector (11) and thrust plate (9) for excessive wear or other damage. Refer to Fig. 92 and the following for specifications of new parts.

Bearing Bore in Housing
(15)................0.7500-0.7505 in.
Bearing (12) Length.....2.425-2.426 in.
Piston Ring Grooves....0.064-0.065 in.
Shaft (17) Concentricity—
Maximum Run-out........0.0006 in.
Shaft (17) Diameter—
Bearing Journal.....0.4400-0.4403 in.
Compressor wheel..0.3123-0.3125 in.
Shaft (17) Shoulder
Length..............2.536-2.537 in.
Thrust Plate (9)
Thickness...........0.107-0.108 in.
Thrust Sleeve (4)
Length..............0.517-0.519 in.
Assembled Bearing Clearances—
Axial (End) Play of
Shaft ....0.002-0.005 in.
Radial Clearance Measured at Exhaust
Turbine Blades ....0.018-0.049 in.
Clearance Between Back of Turbine
Blades and Back Plate (20) Must Be
Even and Within Limits
of ....0.017-0.049 in.

Renew all damaged parts and use new "O" ring (8) and seal rings (5) when reassembling. Use Fig. 92 as a guide when reassembling. Install back plate (20) using new gasket (21). Install seal ring on turbine shaft, lubricate seal ring and install turbine wheel and shaft in bearing housing. Lubricate I.D. and O.D. of bearing (12), install bearing over end of turbine shaft and into bearing housing. Lubricate both sides of thrust plate (9) and install plate (bronze side out) on the aligning dowels. Install oil deflector (11) making certain holes in deflector are positioned over dowel pins. Install new seal ring on spacer sleeve (4), lubricate seal ring and press spacer sleeve into insert (6). Position new "O" ring (8) on insert, lubricate "O" ring and install insert and spacer sleeve assembly in bearing housing, then secure with snap ring (7). Place Compressor wheel on turbine shaft, coat threads and back side of nut (2) with graphite grease, or equivalent, then install and tighten nut to 156 in.-lbs. torque. Apply graphite grease to turbine housing (18), align previously affixed punch marks and tighten screws (24) evenly to 144 in.-lbs. torque. Lock position of screws with lock plate (23). Apply a light coat of graphite grease around machined flange of compressor cover (3). Install compressor cover,

align punch marks and secure cover with cap screws, washer and clamp plates. Tighten cap screws evenly to 60 in.-lbs. torque. Fill the oil inlet with engine oil and turn turbine shaft by hand to lubricate bearing and thrust plate.

Check rotating unit for free rotation within the housings. Cover all openings until the turbocharger is reinstalled.

Use a new gasket and install and prime turbocharger as outlined in paragraph 78.

Turbocharger oil supply pressure at 2200 engine rpm should be within 10 psi of engine oil pressure but never less than 25 psi.

AiResearch Turbocharger

82. **INSPECTION.** Remove turbocharger unit as outlined in paragraph 79. To inspect the removed turbocharger unit, examine turbine wheel and compressor impeller for blade damage, looking through housing openings. Using a dial indicator with plunger extension, check radial bearing play through outlet oil port while moving both ends of turbine shaft equally. Radial bearing clearance should not exceed 0.006 inch. Check shaft end play with dial indicator working from either end. If end play exceeds 0.004 inch or if any of the blades are broken or

damaged, overhaul or renew the turbocharger unit.

83. **OVERHAUL.** Mark across compressor housing (1—Fig. 94), center housing (13) and turbine housing (17) to aid alignment when assembling.

CAUTION: Do not rest weight of any parts on impeller or turbine blades. Weight of only the turbocharger unit is enough to damage the blades.

Remove lock plates and clamp plates (6) from compressor housing (1) and remove housing. Remove lock plates and clamp plates (6) from turbine (17) and remove housing. Hold turbine shaft from turning using the appropriate type of holding fixture for turbine wheel (16) and remove locknut (2).

NOTE: Use a "T" handle or double universal socket to remove locknut in order to prevent bending turbine shaft.

Lift compressor impeller (3) off, then remove center housing from turbine shaft while holding shroud (15) onto center housing. Remove backplate (4), thrust bearing (9) and thrust collar (8). Carefully remove bearing retainers (10) from ends and withdraw bearings (11). Spring (5) is available as an assembly with backplate (4). Refer to Fig. 94A.

CAUTION: Be careful not to damage bearings or surface of center housing

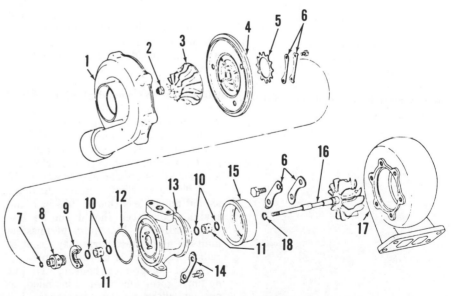

Fig. 94—Exploded view of AiResearch turbocharger assembly used on some models. See Figs. 91 & 92 for Schwitzer models.

1. Compressor housing	6. Clamp plate	11. Bearings (2)
2. Locknut	7. Seal ring	12. "O" ring
3. Compressor impeller	8. Thrust collar	13. Center housing
4. Back plate	9. Thrust bearing	14. Lock plate
5. Spring	10. Bearing retainers	15. Shroud

16. Turbine shaft & wheel
17. Turbine housing
18. Seal ring

when removing retainers. The center two retainers do not have to be removed unless damaged or unseated. Always renew bearing retainers if removed from grooves in housing.

Clean all parts in a cleaning solution which is not harmful to aluminum. A stiff brush and plastic or wood scraper should be used after deposits have softened. When cleaning, use extreme caution to prevent parts from being nicked, scratched or bent.

Inspect bearing bores in center housing (13—Fig. 94) for scored surfaces, out of round or excessive wear. Make certain bore in center housing is not grooved in area where seal (18) rides. Compressor impeller (3) must not show signs of rubbing with either the compressor housing (1) or the backplate (4). Impeller should have 0.0002 inch tight to 0.0004 inch loose fit on turbine shaft. Make certain that impeller blades are not bent, chipped, cracked or eroded. Oil passages in thrust collar (8) must be clean and thrust faces must not be warped or scored. Ring groove shoulders must not have step wear. Inspect turbine shroud (15) for evidence of turbine wheel rubbing. Turbine wheel (16) should not show evidence of rubbing and vanes must not be bent, cracked, nicked, or eroded. Turbine wheel shaft must not show signs of scoring, scratching or overheating. Groove in shaft for seal ring (18) must not be stepped. If turbine shaft journals are damaged, undersized bearings of 0.005 and 0.010 inch may be ordered, and shaft reconditioned if so desired. Check shaft end play and radial clearance when assembling.

If the bearing inner retainers (10) were removed, install new retainers. Oil bearings and install in center housing and install outer retainers. Position the shroud (15) on turbine shaft (16) and install seal ring (18) in groove. Apply a light, even coat of engine oil to

shaft journals, compress seal ring (18) with a thin strong tool such as a dental pick and install center housing (13). Install new seal ring (7) in groove of thrust collar (8), then install thrust bearing so that smooth side of bearing (9) is toward seal ring (7) end of collar. Install thrust bearing and collar assembly over shaft, making certain that pins in center housing engage holes in thrust bearing. Install new rubber seal ring (12), make certain that spring (5) is positioned in backplate (4), then install backplate making certain that seal ring (7) is not damaged. Seal ring will be less likely broken if open end of ring is installed in bore of backplate first. Install lock plates (14) and screws, tightening screws to 75-90 in.-lbs. torque. Install compressor impeller (3) and make certain that impeller is seated against thrust collar (8). Install locknut (2) to 18-20 in.-lbs. torque, then use a "T" handle or double universal joint socket to turn locknut an additional 90° in order to stretch the turbine shaft the necessary 0.0055-0.0065 inch for proper tension.

CAUTION: If "T" handle or double universal joint socket is not used, shaft may be bent when tightening nut (2).

Install turbine housing (17) with clamp plates (6) next to housing, tighten screws to 100-130 in.-lbs., then bend lock plates up around screw heads.

Check shaft end play and radial play at this point of assembly. If shaft end play exceeds 0.004 inch, thrust collar (8) and/or thrust bearing (9) is worn excessively. End play of less than 0.001 inch indicates incomplete cleaning (carbon not all removed) or dirty assembly and unit should be disassembled and cleaned.

If turbine shaft radial play exceeds 0.007 inch, unit should be disassembled and bearings, shaft and/or center housing should be renewed, or shaft should be reconditioned and undersized bearings installed. Center housing bearing bore should not exceed 0.6228 inch and seal bore should not exceed 0.703 inch diameter. Shaft journal diameter should not be less than 0.3994 inch and seal hub diameter should not be less than 0.681 inch. Maximum permissible limits of all of these parts may result in radial play which is not acceptable.

Install compressor housing (1), but do not tighten until unit is installed, so perfect alignment can be made. Fill reservoir with engine oil and protect all openings of turbocharger until unit is installed on tractor. After intake hoses are all assembled, tighten clamp plates (6) to 110-130 in.-lbs. torque to com-

pressor housing (1), and complete the assembly. Prime turbocharger as outlined in paragraph 78.

INTERCOOLER

4630 Model

84. The intake manifold on 4630 model contains an intercooler (3—Fig. 95) to lower the temperature of the intake air. Coolant from the engine cooling system flows through the intercooler core and heat from the compressed (turbocharged) intake air is conducted into the engine coolant, then circulated to the engine radiator where it is permitted to cool (lose heat to the air flowing through the radiator). The intercooler can lower the temperature of the intake air as much as 80-90 degrees F. which will make the intake air more dense and permit more air to be delivered to the engine cylinders. This increased volume of air together with additional quantity of fuel can result in the production of more power from the engine.

Since coolant from the radiator is circulated through the intercooler core, a leak in the core could cause serious damage to the engine by allowing coolant into the combustion area.

To remove the intercooler, drain cooling system, remove muffler, air stack and hood. Remove turbocharger as outlined in paragraph 79. Disconnect both water hoses at intercooler and the ether starting aid pipe. Remove both adapter plates at hose connections to intercooler. Remove the 12 cap screws from underneath intake manifold, lift off manifold cover and intercooler core.

Test and repair procedures are much the same as for a radiator. The inter-

Fig. 94A—AiResearch model back plate and spring must be renewed as an assembly if damaged.

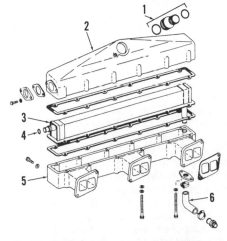

Fig. 95—Exploded view of intake manifold and inter-cooler used on 4630 model.

1. "O" rings and air intake coupling	4. "O" rings
2. Cover	5. Manifold
3. Intercooler core	6. Water inlet hose

cooler can be pressurized with air (20-25 psi) then submerged in water to check for leaks. Repair or renew the aluminum intercooler core as necessary.

Reassemble in reverse order of disassembly using all new gaskets and prime turbocharger as outlined in paragraph 78.

CARBURETOR

4030 and 4230 Gasoline Models

85. The 4030 gasoline tractors are equipped with Marvel-Schebler model TSX-969SL carburetor. The 4230 tractors have a Zenith model 63AW14 (Fig. 96 or Fig. 97). Both models are equipped with an electric fuel shut-off solenoid.

If carburetor has been disassembled make initial adjustments before starting as follows:

4030 (TSX-969SL)
Float level..................¼-inch
(TOP of float to gasket)
Idle air screw........1½ turns open
Load adjusting needle .2½turns open

4230 (63AW14)
Float level..............2-1/64 inch
(Bottom of float to cover-no gasket)
Idle air screw..........1¼ turns open
Load adj. needle......1½-2 turns open
Accelerator pump......3rd notch from
tip on guide rod.

When adjusting the carburetor, first make sure the air cleaner is clean and properly serviced and ignition system is in good condition and correctly timed. Warm the engine to operating temperature and adjust idle speed to 800 rpm. Turn the idle mixture air screw in until engine stumbles and note setting. Turn mixture screw out until engine stumbles again and note number of turns. Final setting of mixture screw should be halfway between the two settings. For best results the load needle should be adjusted under operating conditions for best power and economy. A pto dynamometer is recommended for this adjustment.

GASOLINE FUEL PUMP AND FILTERS

86. A sediment bowl is located at the rear of gasoline tank, and should be removed and strainer cleaned with each tune up, or when water is observed in bowl.

An in-line gasoline filter is located between the fuel pump and carburetor. Renew the filter at each tune up, and note arrow, which indicates direction of flow. If water is found in system, it may also be necessary to drain the gasoline tank and carburetor; and clean sediment bowl.

The gasoline fuel pump is camshaft operated, and must be renewed if found to be faulty. Before renewing pump, check fuel pressure, which should be 3½-4½ psi at slow idle, and be sure that sediment bowl or line to pump is not restricted.

DIESEL FUEL SYSTEM

FUEL FILTER, PRIMARY PUMP AND LINES

All Models

90. Models without turbochargers are equipped with a single two-stage fuel filter and sediment bowl, and turbocharged models have dual filters. On all models the bowl assembly should be renewed when necessary. The primary fuel pumps on 4030 and 4230 models have no sediment bowl, with all filtering done at the single two-stage filter. Turbocharged 4430 and 4630 models use a fuel transfer pump between the tank and dual two-stage filters, that has a glass sediment bowl with a renewable filter.

The fuel tank is mounted vertically in front of radiator, and shut-off valve on all models is accessible after removing right grille screen. The fuel

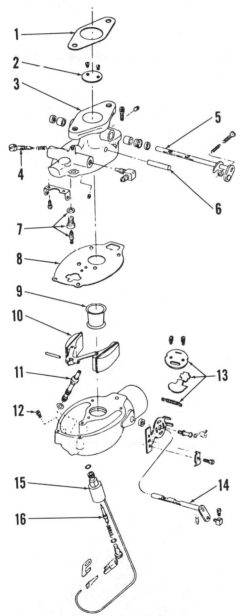

Fig. 96—Exploded view of Marvel-Schebler carburetor used on 4030 gasoline tractors.

1. Gasket
2. Throttle plate
3. Cover
4. Idle needle
5. Throttle shaft
6. Spring pin
7. Needle & seat
8. Cover gasket
9. Venturi
10. Float
11. Discharge nozzle
12. Power jet
13. Choke assembly
14. Choke shaft
15. Shut-off solenoid
16. Load needle

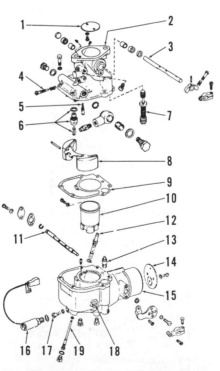

Fig. 97—Exploded view of Zenith carburetor used on 4230 gasoline tractors.

1. Throttle plate
2. Cover
3. Throttle shaft
4. Idle needle
5. Idle jet
6. Needle & seat
7. Accelerator pump
8. Float
9. Cover gasket
10. Venturi
11. Choke shaft
12. Discharge nozzle
13. Power valve
14. Choke plate
15. Bowl
16. Solenoid & load needle
17. Main jet
18. Pump check valve
19. Power jet

tank drain valve is mounted in the lowest part of fuel tank, forward of the front axle.

91. BLEEDING. To bleed the system, refer to Fig. 100. Open filter bleed plug on all models. On models without turbocharger, actuate hand primer lever on fuel pump until air-free fuel flows out bleed plug opening.

NOTE: If no resistance is felt when pumping lever, turn engine so that the camshaft is not on its pump stroke.

One bleeding operation is usually all that is required. Tighten bleed plug. Make sure hand primer lever is in down position before attempting to start engine.

On models equipped with turbocharger, the fuel transfer pump is attached to the injector pump (Fig. 101) and the hand primer must be unscrewed until it can be pulled up. Pump the primer until air-free fuel flows out the open bleed plug hole.

Loosen the pressure line connections at injector assemblies and, with throttle open, turn engine over with starter until fuel flows from all injector lines. Tighten the connections and start engine. If engine will not start or misses, repeat the above procedure until system is free of trapped air.

FUEL TRANSFER PUMP

All Models

92. The fuel transfer pump on models without turbocharger (4030 and 4230) is mounted on right side of engine block, driven by the camshaft, and includes a hand primer lever. This type of fuel pump has no sediment bowl and is available as an assembly only.

INJECTOR NOZZLES

4030 and 4230 (Before Eng. Serial No. 500000)

Roosa-Master pencil injector nozzles are used on these models. Two different types are used and should not be interchanged. Refer to Fig. 102 and Fig. 103 for identification.

94. TESTING AND LOCATING A FAULTY NOZZLE. If rough or uneven engine operation or misfiring indicate a faulty injector, the defective unit can usually be located as follows:

With engine running at the speed where malfunction is most noticeable (usually slow idle speed), loosen the compression nut on high pressure line for each injector in turn, and listen for

a change in engine performance. As in checking spark plugs, the faulty unit is the one which, when its line is loosened, least affects the running of the engine.

If a faulty nozzle is found and considerable time has elapsed since the injectors have been serviced, it is recommended that all nozzles be removed and checked, or that new or reconditioned units be installed. Refer to the following paragraphs for removal and test procedure.

95. REMOVE AND REINSTALL. Wash injector, lines and surrounding area with clean diesel fuel to remove any accumulation of dirt or foreign material. Remove air stack, muffler and hood. Disconnect leak-off pipe at return line fitting and at injection pump.

NOTE: If working on a nozzle near alternator, disconnect battery ground cables to prevent shorting a tool against alternator terminal.

Expand lower clamp on each leak-off boot and move clamp upward next

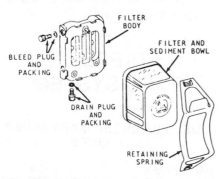

Fig. 100—Exploded view of single renewable two-stage filter and sediment bowl element used on 4030 and 4230 models. Models with turbocharger are equipped with dual, two-stage filters.

Fig. 101—Fuel transfer pump with hand primer as used on turbocharged models.

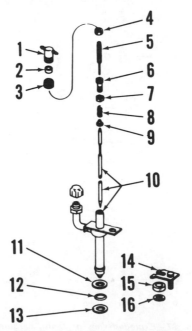

Fig. 102—Exploded view of Roosa Master injector used on 4030 model.

1. Leak-off cap
2. Grommet
3. Nut
4. Lift screw nut
5. Lift screw
6. Pressure screw
7. Nut
8. Spring
9. Spring seat
10. Valve & body assembly
11. Upper washer
12. Nozzle seal
13. Lower washer
14. Clamp
15. Spacer
16. Washer

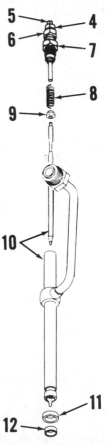

Fig. 103—Exploded view of Roosa Master injector used on 4230 models before engine serial number 500000.

4. Lift screw nut
5. Lift screw
6. Pressure screw
7. Nut
8. Spring
9. Spring seat
10. Valve & body assembly
11. Washer
12. Seal

to top clamp; then remove leak-off pipe and all boots as a unit.

Disconnect high-pressure line, remove nozzle clamp cap screw, clamp and spacer; then withdraw injector assembly.

NOTE: If injector cannot be easily withdrawn by hand, the special OTC puller, JDE-38 will be required. DO NOT attempt to pry nozzle from its bore.

Before reinstalling the injector nozzle clean nozzle bore in cylinder head using OTC Tool JDE-39, then blow out foreign material with compressed air. Turn tool clockwise only when cleaning nozzle bore. Reverse rotation will dull tool.

Renew carbon seal at tip of injector body and seal washer at upper seat whenever injector has been removed. The protector cap can be used to push seal onto nozzle body, if cap is available.

NOTE: Nozzle tip may be cleaned of loose or flaky carbon using a brass wire brush. DO NOT use a brush, scraper or other abrasive on Teflon coated surface of nozzle body between the seals. The coating may become discolored by use, but discoloration is not harmful.

Insert the dry injector nozzle in its bore using a twisting motion. Tighten pressure line connection finger tight; then install hold-down clamp, spacer and cap screw. Tighten cap screw to a torque of 20 ft.-lbs. Bleed the injector if necessary, as outlined in paragraph 91, then tighten pressure line connection to approximately 35 ft.-lbs. Complete the assembly by reversing disassembly procedure.

96. **NOZZLE TEST.** A complete job of testing and adjusting an injector requires the use of special test equipment. Only clean approved testing oil should be used in tester tank. The nozzle should be tested for opening pressure, seat leakage, back leakage and spray pattern. When tested, the nozzle should open with a sharp popping or buzzing sound and cut off quickly at end of injection with a minimum of seat leakage and a controlled amount of back leakage.

Use the tester to check injector as outlined in the following paragraphs:

CAUTION: Fuel leaves the nozzle tip with sufficient force to penetrate the skin. Keep unprotected parts of body clear of nozzle spray when testing.

97. OPENING PRESSURE. Before conducting the test, operate tester lever until fuel flows, then attach the injector using No. 16492 Special Adapt-

er. Close the valve to tester gage and pump the tester lever a few quick strokes to be sure nozzle valve is not plugged, that all spray holes are open and that possibilities are good that injector can be returned to service without overhaul.

Open valve to tester gage and operate tester lever slowly while observing gage reading. Opening pressure should be 3000 psi, if it is not, adjust opening pressure and valve lift as follows:

Loosen clamp nut (4—Fig. 102 or Fig. 103) while holding pressure adjusting screw (6) from turning, then back out lift adjusting screw (5) at least two turns to insure against bottoming. Turn adjusting screw (6) until specified opening pressure is obtained. While holding adjusting screw from turning and before tightening clamp nut, turn lift adjusting screw until it bottoms; then back out ½-turn on 4030 models, and ¾-turn on 4230 models. Tighten clamp screw and recheck opening pressure. Reinstall as outlined in paragraph 95, and bleed system as outlined in paragraph 91.

NOTE: When adjusting a new injector or an overhauled injector with a new pressure spring, set the pressure at 3200 psi to allow for initial pressure loss as the spring takes a set.

98. SPRAY PATTERN. The finely atomized nozzle spray should be evenly distributed around the nozzle. Check for clogged or partially clogged orifices or for a wet spray which would indicate a sticking or improperly seating nozzle valve. If the spray pattern is not satisfactory, disassemble and overhaul the injector as outlined in paragraph 101.

99. SEAT LEAKAGE. Pump the tester handle slowly to maintain a gage pressure of 2700 psi while examining nozzle tip for fuel accumulation. If nozzle is in good condition, there should be no noticeable accumulation for a period of at least 10 seconds. If a drop or undue wetness appears on nozzle tip, renew the injector or overhaul as outlined in paragraph 101.

100. BACK LEAKAGE. Loosen compression nut and reposition nozzle so that spray tip is slightly higher than adjusting screw end of nozzle, then maintain a gage pressure of 1500 psi. After the first drop falls from adjusting screw, leakage should be at the rate of 3-10 drops in 30 seconds. If leakage is excessive, renew the injector.

101. **OVERHAUL.** First clean outside of injector thoroughly. Place nozzle in a holding fixture and clamp the fixture in a vise. NEVER tighten vise jaws on

nozzle body without the fixture. Refer to Fig. 102 or Fig. 103. Loosen locknut (4) and back out pressure adjusting screw (6) containing lift adjusting screw (5). Slip nozzle body from fixture, invert the body and allow spring seat (9) and spring (8) to fall from nozzle body into your hand. Catch nozzle valve (10) by its stem as it slides from body. If nozzle valve will not slide from body, use the special retractor (16481) or discard the injector assembly.

Nozzle valve and body are a matched set and should never be intermixed. Keep parts for each injector separate and immerse in clean diesel fuel in a compartmented pan as injector is disassembled.

Clean all parts thoroughly in clean diesel fuel using a brass wire brush. Hard carbon or varnish can be loosened with a suitable non-corrosive solvent.

NOTE: Never use a steel wire brush or emery cloth on spray tip.

Clean the spray tip orifices using the appropriate size cleaning needle. 4030 nozzles have 0.011 inch diameter orifices, 4230 nozzles have 0.012 inch orifices.

Clean the valve seat using a Valve Tip Scraper and light pressure. Use a Sac Hole Drill to remove carbon from inside of tip.

Piston area of valve can be lightly polished by hand if necessary, using Roosa Master No. 16489 lapping compound. Use the valve retractor to turn valve. Move valve in and out slightly while turning but do not apply down pressure while valve tip is in contact with seat.

Valve and seat are ground to a slight interference angle. Seating areas may be cleaned up if necessary using a small amount of 16489 lapping compound, very light pressure and no more than 3 to 5 turns of valve on seat. Thoroughly flush all compound from valve body after polishing.

When assembling, back out lift adjusting screw (5), and reverse the disassembly procedure using Fig. 102 or Fig. 103 as a guide. Adjust opening pressure and valve lift as outlined in paragraph 97 after valve is assembled.

4430 (Before Eng. Serial No. 445570) and 4630 (Before Eng. Serial No. 445948)

Early 4430 and 4630 models are equipped with Robert Bosch KDL (21mm) nozzles. Refer to Fig. 104. Later models use KDEL nozzles.

102. **REMOVE AND REINSTALL.** Refer to paragraph 94 for testing pro-

cedures to determine if injector nozzle shows indication of a malfunction before removing for service.

If nozzle to be removed is near the alternator, disconnect battery ground strap to prevent a short circuit through tools. Wash injector and surrounding area with clean diesel fuel, disconnect leak-off line and fuel pressure line and use special tool (JDE-69 or JDE-69A) to remove gland nut (9—Fig. 104). The gland nut will raise the nozzle out of cylinder head as it is removed.

When reinstalling, make sure injector and hole in cylinder head are clean and dry. Nozzle seat reamer (JDE-99) can be used to clean nozzle seat in head. Threads in head for gland nut (9) can be cleaned using a metric (M24 x 1.5) tap. This is different thread than later models with KDEL nozzles. Renew nozzle gasket if necessary. Apply anti-seize compound to bottom, inner bore and threads of gland nut (9), then tighten gland nut to 35 to 45 ft.-lbs. Renew leak-off line gaskets if necessary, hold fitting and tighten 12 millimeter head screw to 20 ft.-lbs. Bleed injectors as outlined in paragraph 91 and tighten pressure lines to 35 ft.-lbs. torque.

103. **NOZZLE TEST.** A complete job of testing and adjusting an injector requires the use of special test equipment. Only clean approved testing oil should be used in tester tank. The nozzle should be tested for opening pressure, seat leakage and spray pattern. When tested, the nozzle should open with a soft chatter, and then only when the lever is moved very rapidly. A bent or binding nozzle valve can prevent chatter. Spray will be broad and well atomized if injector is working properly.

Use the tester to check injector as outlined in the following paragraphs:

CAUTION: Fuel leaves the nozzle tip with sufficient force to penetrate the skin. Keep unprotected parts of body clear of nozzle spray when testing.

104. OPENING PRESSURE. Before conducting the test, operate tester lever until fuel flows, then attach the injector using the proper adapter. Close the valve to tester gage and pump the tester lever a few quick strokes to be sure nozzle valve is not plugged, that all spray holes are open and that possibilities are good that injector can be returned to service without overhaul.

Open valve to tester gage and operate tester lever slowly while observing gage reading. Opening pressure should be 3100 psi; if it is not, re-

check by releasing pressure and retesting. If pressure is still not correct, remove plug (3—Fig. 104) and change shim (4) until opening pressure is correct. Use only specially hardened shims. Shims are available in 0.002 inch steps from 0.043 to 0.059 inch. Each 0.002 inch increase in shim thickness varies the pressure by about 75 psi. Opening pressure should not vary more than 50 psi between nozzles. If pressure is not correct after changing shims, disassemble injector and recondition.

NOTE: When adjusting a new injector or an overhauled injector with a new pressure spring, set the pressure at 3200 to 3350 psi to allow for initial pressure loss as the spring takes a set.

105. SPRAY PATTERN. The finely atomized nozzle spray should be evenly distributed around the nozzle. Check for clogged or partially clogged orifices.

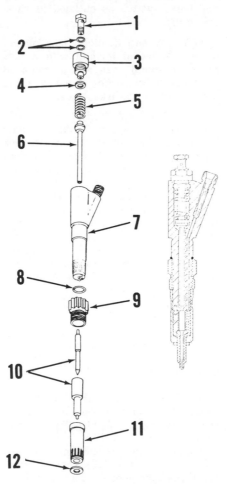

Fig. 104—Exploded view of Robert Bosch KDL injector nozzle used on early 4430 and early 4630 models.

1. Screw	7. Body
2. Gasket (2)	8. "O" ring
3. Plug	9. Gland nut
4. Adjusting shim	10. Nozzle
5. Spring	11. Retaining nut
6. Spindle	12. Washer

Nozzles for 4430 have four orifices and nozzles for 4630 models have three. Also check for a wet spray which would indicate a sticking or improperly seating nozzle valve. If the spray pattern is not broad and even, and very rapid stroking of the tester handle does not cause injector to chatter softly, disassemble and overhaul the injector as outlined in paragraph 107.

106. SEAT LEAKAGE. Pump the tester handle slowly to maintain a gage pressure of 285 psi while examining nozzle tip for fuel accumulation. If nozzle is in good condition, there should be no noticeable accumulation for a period of at least 10 seconds. If a drop or undue wetness appears on nozzle tip, renew the injector or overhaul as outlined in paragraph 107.

107. **OVERHAUL.** First clean outside of injector thoroughly and place in a soft jawed vise. Remove plug (3—Fig. 104), then carefully withdraw shim (4), spring (5) and spindle (6). Turn nozzle body over in vise and remove nozzle retaining nut (11) with a box end wrench. DO NOT use a pipewrench. Remove nozzle valve assembly (10) and reinstall retaining nut on body (7) to protect the lapped end surface. If nozzle valve cannot be removed easily, soak assembly in carburetor cleaner, acetone or other commercial solvent intended to free stuck valves. Use care to keep parts clean and free from grit by submerging in a pan of clean diesel fuel, and handle only with hands that are wet with fuel. Avoid mixing of parts with another injector, and do not allow any lapped surface to come in contact with a hard object.

Valves should be cleaned of all carbon and washed in diesel fuel. Hard carbon may be cleaned off with a brass wire brush. NEVER use a steel wire brush or emery cloth on valve or tip. Use a cleaning wire 0.003 to 0.004 inch smaller than nozzle orifices to clean nozzle tips. The number and size of orifices are etched on nozzle tip such as "4 x 0,33". The "4" indicates four orifices, "0,33" indicates that each is 0.33 mm diameter. The following data applies to standard nozzles on models with KDL nozzles.

4430 (prior to Eng. Serial No. 445570) . . . . 4 x 0,33
    Orifice diameter . . . . 0.013 in.
    Opening pressure—New 3200-3350 psi.
      Used . . . . 3100 psi.
4630 (prior to Eng. Serial No. 335846) . . . . 4 x 0,33
    Opening pressure—
      New . . . . 3200-3350 psi.
      Used . . . . 3100 psi.

4630 (Eng. Serial No's. 335846-
   445947) ....3 x 0,40
   Orifice diameter ....0.0157 in.
   Opening pressure—
      New............3200-3350 psi.
      Used ....3100 psi.

A stone may be used to cut flat surfaces on two sizes of cleaning wire to
aid in removing carbon from orifices.
Finish cleaning orifices by using a wire
0.001-inch smaller than hole diameter.
A pin vise should be used to hold
cleaning wires and wire should extend
only about 1/32-inch from vise to
prevent breakage. Clean seat in nozzle
(10) with sac hole drill furnished with
cleaning kit. When held vertically, a
valve that is wet with fuel should slide
down to the seat in nozzle under its
own weight.

Inspect all lapped and seating surfaces for excessive wear or damage.
Check spindle, spring, shims and seats.
Renew any parts in question and replace shims only if they are smooth and
flat. Edge type filter in fuel inlet
passage of body (7) can be cleaned by
blowing air through passage from
nozzle end of body. This will provide a
reverse flushing action in filter.

Assemble in reverse order of disassembly. Submerge valve and nozzle in
fuel while assembling, and make sure
all other parts are wet with fuel. Do
not dry parts with air or towels before
assembly. With injector body clamped
in a soft jawed vise, tighten screw plug
(3) to 36-44 ft.-lbs. and retaining nut
(11) to 44-58 ft.-lbs. Retest injector as
outlined in paragraphs 103, 104, 105
and 106. Use a new gasket when reinstalling injector in engine. Bleed system as outlined in paragraph 91.

4230 (After Eng. Serial No. 499999), 4430 (After Eng. Serial No. 445569), and 4630 (After Eng. Serial No. 445947) Models

These late models are equipped with
Robert Bosch KDEL injector nozzles.
Refer to Fig. 105 for identifying fea-
tures. All threaded components are
metric sizes.

108. **REMOVE AND REINSTALL.**
Refer to paragraph 94 for testing procedures to determine if injector nozzle
shows indication of a malfunction before removing for service.

If nozzle to be removed is near the
alternator, disconnect battery ground
strap to prevent a short circuit
through tools. Wash injector and surrounding area with clean diesel fuel,
disconnect leak-off line and fuel pressure line and use special tool (JDE-92)

to remove gland nut (9—Fig. 105). The
gland nut will raise the nozzle out of
cylinder head as it is removed.

When reinstalling, make sure injector and hole in cylinder head are clean
and dry. Nozzle seat reamer (JDE-99)
can be used to clean nozzle seat in
head. Threads in head for gland nut (9)
can be cleaned using a metric (M28 x
1.5) tap. This thread is different than
earlier models with KDL nozzles.
Threads for leak-off connectors are
metric (M6 x 1). Apply anti-seize compound to bottom, inner bore and
threads of gland nut and to nozzle
barrel before installing.

Renew nozzle gasket (12) if necessary. Use special tool JDE-92, align
leak-off threaded port and tighten
gland nut to 55-65 ft.-lbs. torque. Install
leak-off connectors, attach leak-off line
and connect delivery pipes to nozzles.
Tighten fittings between delivery pipes
and nozzles to 35 ft.-lbs. torque. Bleed
injectors as outlined in paragraph 91.

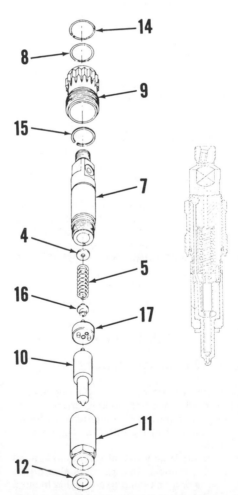

*Fig. 105—Exploded view of 21mm Robert
Bosch KDEL injector typical of type used on
Late 4230, Late 4430, Late 4630 models.*

109. **NOZZLE TEST.** A complete job
of testing and adjusting an injector requires the use of special test equipment. Only clean approved testing oil
should be used in tester tank. The nozzle should be tested for opening pressure, seat leakage and spray pattern.
When tested, the nozzle should open
with a soft chatter, and then only when
the lever is moved very rapidly. A bent
or binding nozzle valve can prevent
chatter. Spray will be broad and well
atomized if injector is working properly.

Use the tester to check injector as
outlined in the following paragraphs:

CAUTION: Fuel leaves the nozzle tip
with sufficient force to penetrate the
skin. Keep unprotected parts of body
clear of nozzle spray when testing.

110. OPENING PRESSURE. Before
conducting the test, operate tester
lever until fuel flows, then attach the
injector using the proper adapter.
Close the valve to tester gage and
pump the tester lever a few quick
strokes to be sure nozzle valve is not
plugged, that all spray holes are open
and that possibilities are good that injector can be returned to service
without overhaul.

Open valve to tester gage and
operate tester lever slowly while observing gage reading. Opening pressure should be 3200 psi for 4040, 4230
and 4240 models; 3800 psi for all turbocharged models. If pressure is too low,
recheck by releasing pressure and
retesting. If pressure is still not correct, refer to paragraph 113, disassemble nozzle and change thickness of shim
(4—Fig. 105) until opening pressure is
correct. Use only specially hardened
shims. Shims are available in 0.002 inch
steps from 0.039 to 0.077 inch. Each
0.002 inch increase in shim thickness
varies the pressure by about 100 psi.
Opening pressure should not vary more
than 50 psi between nozzles. If pressure is not correct after changing
shims, disassemble injector and recondition.

NOTE: When adjusting a new injector
or an overhauled injector with a new
pressure spring, set the pressure at 3350
psi for 4230 models; to 4050 psi for turbocharged models. The increase in pressure
is to allow for initial pressure loss as the
spring takes a set.

111. SPRAY PATTERN. The finely
atomized nozzle spray should be evenly
distributed around the nozzle. Check
for clogged or partially clogged orifices
or for a wet spray which would indicate
a sticking or improperly seating nozzle

valve. If the spray pattern is not broad and even, and very rapid stroking of the tester handle does not cause injector to chatter softly, disassemble and overhaul the injector as outlined in paragraph 113.

112. SEAT LEAKAGE. Pump the tester handle slowly to maintain a gage pressure of 285 psi while examining nozzle tip for fuel accumulation. If nozzle is in good condition, there should be no noticeable accumulation for a period of at least 10 seconds. If a drop or undue wetness appears on nozzle tip, renew the injector or overhaul as outlined in paragraph 113.

113. OVERHAUL. First clean outside of injector thoroughly and clamp flats of nozzle retaining nut (11—Fig. 105) in a soft jawed vise. If not already removed, the leak-off connector fitting must be removed before removing upper snap ring (14), "O" ring (8), gland nut (9) and lower snap ring (15). Clamp the flats at fuel inlet end of nozzle holder (7) in a soft jawed vise, then unscrew the nozzle retaining nut (11). Be careful not to mar the polished surfaces of holder (7), intermediate plate (17) or nozzle (10). It may be necessary to soak nozzle assembly in carburetor cleaner, acetone or other commercial solvent if the valve does not fall freely from nozzle (10).

Clean all parts, then use care to keep parts clean and free from grit by submerging in a pan of clean diesel fuel, and handle only with hands that are wet with fuel. Avoid mixing of parts and other injector, and do not allow any lapped surface to come in contact with a hard object.

Valves should be cleaned of all carbon and washed in diesel fuel. Hard carbon may be cleaned off with a brass wire brush. NEVER use a steel wire brush or emery cloth on valve or tip. Use a cleaning wire 0.003 to 0.004 inch smaller than nozzle orifices to clean nozzle tips. The number and size of orifices are etched on nozzle tip such as "4 x 0,33". The "4" indicates four orifices, "0,33" indicates that each is 0.33 mm diameter. The following data applies to standard nozzles on models with KDEL nozzles.

4230 (After Eng. Serial No.
499999)..................5 x 0,25
Orifice diameter..........0.010 in.
Opening pressure—New.....3350 psi
Used..................3200 psi

4430 (After Eng. Serial No.
445569)..................4 x 0,33
Orifice diameter..........0.013 in.

Opening pressure—New.....4050 psi
Used..................3800 psi
4630 (After Eng. Serial No.
445947)..................4 x 0,345
Orifice diameter..........0.0136 in.
Opening pressure—New.....4050 psi
Used..................3800 psi

A stone may be used to cut flat surfaces on two sides of cleaning wire to aid in removing carbon from orifices. Finish cleaning orifices by using a wire that is 0.001 inch smaller than hole diameter. A pin vise should be used to hold cleaning wires and wire should extend only about 1/32-inch from vise to prevent breakage. Clean seat in nozzle (10) with sac hole drill furnished with cleaning kit. When held vertically, a valve that is wet with fuel should slide down to the seat in nozzle under its own weight.

Inspect all lapped and seating surfaces for excessive wear or damage. Check spindle, spring, shims and seats. Renew any parts in question and replace shims only if they are smooth and flat. Edge type filter in fuel inlet passage of body (7) can be cleaned by blowing air through passage from nozzle end of body. This will provide a reverse flushing action in filter.

Assemble in reverse order of disassembly. Submerge valve and nozzle in fuel while assembling, and make sure all other parts are wet with fuel. Do not dry parts with air or towels before assembly.

Apply anti-seize compound to bottom, inner bore and to threads of gland nut (9) and to body (7) before assembling snap rings (14 & 15), gland nut (9) and "O" ring (8). Clamp body in a soft jawed vise, then assemble shim (4), spring (5), spring seat (16), intermediate plate (17), nozzle (10) and retaining nut (11). Tighten retaining nut (11) to 44-58 ft.-lbs. torque, then retest injector as outlined in paragraphs 109, 110, 111 and 112. Use a new gasket (12) when reinstalling in engine. Bleed system as outlined in paragraph 91.

INJECTION PUMP

Roosa-Master Pumps

Roosa-Master model JDB pump is used on 4030 model. The pump, driven by the engine timing gears, is located on left side of engine and mounted on engine front plate.

Roosa-Master model DM pump is used on 4230 model. This pump is also driven by the engine timing gears, but is located on right side of engine and is attached to a mounting flange which is part of the engine block casting.

On all models, proper injection timing depends upon correct positioning of timing marks on pump and engine as outlined in paragraphs 115, 116 and 117.

Service to the injection pump requires use of special tools and specialized training which is beyond the scope of this manual. This service section will cover only the information required for removal, installation and field adjustments of the pump.

115. REMOVE AND REINSTALL. To remove the Roosa-Master fuel injection pump, first shut off fuel supply and thoroughly clean dirt from pump, lines and connections.

NOTE: Do not steam clean or pour water on a pump while it is warm or running, as this could cause pump to seize.

Remove clamps on injector lines, six connectors at pump and loosen lines at injector nozzles to avoid bending lines. Plug opening in fuel lines to prevent dirt from entering. Remove line from filter and fuel leak-off line.

On all except 4030 models, remove the crankcase vent hose and clean area around timing gear access cover at front of timing gear cover. Remove access cover, then remove the three screws which attach pump drive gear to drive hub. Clean the area around the fuel transfer pump, then disconnect fuel lines and remove transfer pump.

On all models, remove cover from timing window, install engine rotating tool and rotate engine until marks on governor weight retainer and cam ring are aligned as shown in Fig. 109. Disconnect speed control rod and fuel shut-off cable. Remove the pump mounting stud nuts, then lift pump back away from mounting. There are two mounting stud nuts on 4030 model; three stud nuts on 4230 model. The drive gear and shaft will remain in engine on 4030 models.

To reinstall pump, reverse removal procedure. Make sure that No. 1 piston is at TDC on compression stroke and that the timing marks on cam ring and governor weight are aligned. Renew pump mounting seal if necessary. On 4030 models, use JD 256 or No. 13371 installation tool to install new seals on pump shaft. Be sure that reference mark on tang of pump shaft aligns with mark inside pump. This will prevent timing from being 180° off. The backlash in gears can allow pump timing to be off several degrees, so it is very important to recheck timing with engine running as outlined in paragraph 117. Bleed fuel system as outlined in paragraph 91. If pump drive gear must be renewed, remove timing cover (paragraph 56).

116. STATIC TIMING. To check injection pump static timing, proceed as outlined in paragraph 115 but without removing pump.

If adjustment is required, loosen mounting stud nuts and turn pump body until mark on cam ring aligns with mark on governor weight retainer (Fig. 109). Turn engine in normal direction two complete turns and recheck setting.

117. ADVANCE TIMING. The injection pump has automatic speed advance which is factory set and will not normally need to be checked or reset. Minor adjustments can, however, be made without removal or disassembly of the pump; proceed as follows:

Shut off fuel, remove timing window cover (Fig. 109) and install timing window JD 259 or No. 13366 on 4030 models, or No. 19918 for other models. Make sure that pump has been static timed to engine as outlined in paragraph 116, and engine is up to operating temperature. Install a master

Fig. 107—View of Roosa Master JDB pump installed on 4030 tractor. Refer to Fig. 108 for installed view of DM Roosa Master pump typical installation.

Fig. 108—View of DM Roosa Master injection pump, typical of all models equipped with this pump.

tachometer on speed-hour meter drive using adapter JDE-28. Each mark on timing window is 2 degrees. Timing may be changed by removing seal cap on advance trimmer screw, loosening locknut, and with engine running, turning trimmer screw out to advance, or in to retard pump timing on 4030 models, or turning trimmer screw IN to advance, or OUT to retard on all except 4030 models. (Both pumps turn the same direction, but the trimmer screws are on opposite sides of pump). Specifications are as follows:

4030 [JDB Pump]
1150 rpm (NO LOAD) . . . . . . 4° advance
1500 rpm (FULL LOAD) . . . 4° advance
2300 rpm (FULL LOAD) . . . 6° advance
4230 [Pump DM2633]
1300 rpm (NO LOAD) . . . . . . 5° advance
1700 rpm (FULL LOAD) . . . 5° advance
2100 rpm (FULL LOAD) . . . 6° advance
4230 [Pump DM4629]
2100 rpm (FULL LOAD) . . . 9° advance
Total Advance—FULL
LOAD . . . . 9° advance
NO LOAD . . . . 9° advance

Full load must be set using a pto dynamometer, but if a dynamometer is not available, the no load advance setting can be used, which would be the next best way to adjust the pump.

If proper advance cannot be obtained, pump must be either removed and adjusted on a test stand, overhauled or renewed.

Robert Bosch Pumps

Robert Bosch multiple plunger injection pumps are used on turbocharged models. Series A-2000 pumps are used on 4430 tractors; series A-3000 pumps are used on 4630 tractors.

Series A-2000 and A-3000 pumps are multiple plunger in-line pumps, with a governor and an aneroid control. They are equipped with an externally mounted fuel transfer pump which includes a hand primer pump, and are lubricated by engine oil pressure.

Series A-3000 pump has larger diameter plungers and other parts, than the smaller A-2000 pumps, in order to increase the amount of fuel delivered to the injectors.

Service to the injection pump requires use of special tools and specialized training which is beyond the scope of this manual. This service section will cover only the information required for removal, installation and field adjustments of the pump.

119. REMOVE AND REINSTALL. To remove the Robert Bosch multiple plunger fuel injection pump, first shut off fuel supply and thoroughly clean dirt from pump, lines and connections.

NOTE: Do not steam clean or pour water on a pump while it is warm or running, as this could cause pump to seize.

Cap all fittings as they are disconnected to prevent dirt entry. Remove access plate from front of timing gear cover and remove timing hole plug which is just to the rear of engine oil filler cap. Install engine rotation tool and timing pin in flywheel housing and refer to Fig. 112. Rotate engine in normal rotation direction until the mark on pump drive hub lines up with pointer mark, and timing pin enters hole in flywheel. Remove the three cap screws holding pump drive gear to pump. Disconnect the fuel and oil lines to pump and the pipe to aneroid. Disconnect the injector lines, return hose, speed control rod and fuel shut-off

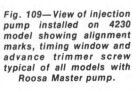

Fig. 109—View of injection pump installed on 4230 model showing alignment marks, timing window and advance trimmer screw typical of all models with Roosa Master pump.

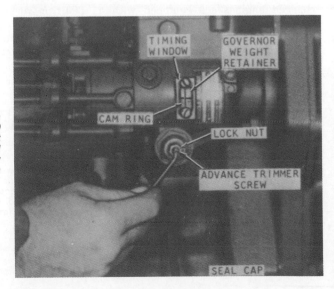

cable. It may be necessary on some models to unbolt and remove oil filter and valve body from engine block. On all models, remove the mounting nuts which hold pump to engine and withdraw pump.

When installing injection pump, No. 1 piston must be at TDC on compression stroke and pump timing marks aligned (Fig. 112) for proper timing. Compression stroke can be found with rocker cover removed, by turning engine until No. 1 intake valve is closing and continuing to turn until timing pin enters hole in flywheel. Renew "O" ring on front bearing plate on pump, lubricate liberally and slide pump onto mounting studs. Make sure that the three drive gear slots are nearly centered with pump drive holes and drive hub mark is aligned with pointer mark before tightening the three cap screws. Tighten mounting stud nuts and drive gear cap screws to 35 ft.-lbs. torque. Reverse disassembly procedure and add 1 U.S. pint of engine oil through filler plug in side cover of pump. Check timing as outlined in paragraph 120 and bleed system as outlined in paragraph 91.

120. TIMING. To check injection pump static timing, proceed as outlined in paragraph 119 but without removing pump, lines or linkage.

If adjustment is required, loosen the three cap screws on pump drive gear until the drive hub can be moved in the slotted holes in the drive gear. When timing marks are aligned (Fig. 112) retighten three cap screws to 35 ft.-lbs. torque, rotate engine two complete revolutions and recheck timing marks and condition of drive gear teeth. No further timing checks are necessary with this type of pump.

121. ANEROID. Tractors with turbocharged engines are equipped with a diaphragm type control unit, which fits on top of governor assembly on the Robert Bosch fuel injection pump (Fig. 113). This diaphragm is operated by positive pressure from the intake manifold, which results from the turbocharger producing boost pressure on a hard pull. Until positive pressure is built up by the turbocharger, the vertical shaft which extends down into governor housing provides a stop to limit the travel of the fuel control rack. This allows the engine to accelerate without producing black smoke unnecessarily. When the turbocharger builds sufficient pressure to depress the diaphragm spring in the aneroid, the aneroid shaft moves down and allows the fuel control rack to move farther open, which delivers more fuel at the time the engine can burn it.

Aneroids must be adjusted to the pump on a test stand, so if it becomes necessary to renew the diaphragm or the entire assembly, pump should be recalibrated on a test stand.

THROTTLE LINKAGE

Gasoline Models

124. LINKAGE AND SPEED ADJUSTMENT. Recommended idle and loaded engine speeds are as follows:

	Idle Speed	Loaded Speed
4030 (Full throttle)	2700	2500
Slow idle	800	
4230 (Full throttle)	2400	2200
Slow idle	800	

With throttle lever in the wide open position, adjust the front control rod nuts to obtain high idle speed as listed in preceding table. Move throttle lever back until 800 rpm slow idle speed is reached. Adjust lever stop screw (Fig. 115) until slow idle always returns to 800 rpm when lever is moved to slow idle position against stop screw, then tighten jam nut. On 4230 models only, adjust governor arm stop screw (inset) until there is a 1/32-inch gap between head of stop screw and anti-surge leaf spring. If surging occurs during operation, this 1/32-inch gap may be closed up until surging stops. Be sure 800 rpm slow idle can still be obtained after this adjustment is made. Adjust the lever friction spring screw until approximately 8 lbs. of force is required to change lever position.

Diesel Models

125. LINKAGE AND SPEED ADJUSTMENT. Fast idle speed should be 2660 rpm for 4030 models; 2380-2420 rpm for 4230 models; 2325-2425 rpm for all turbocharged models. Slow idle speed on all models should be set at 780-820 rpm. To adjust, refer to Fig. 116, then proceed as follows:

Check the force necessary to move the hand throttle lever. If force necessary is not approximately 8 lbs., remove access panel and turn the lever friction spring screw.

Move throttle lever all the way forward until it stops. Adjust the front control rod by loosening nuts on ball joint connections, or the turnbuckle if so equipped. The arm on injector pump

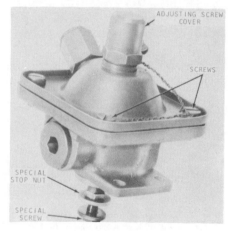

Fig. 113—Aneroid Control unit typical of type used on Robert Bosch injector pumps. Refer to paragraph 121 for explanation of operation.

Fig. 110—View of A-Series Robert Bosch injection pump typical of installation on 4430 and 4630 models.

Fig. 112—View of Robert Bosch injection pump drive gear and timing marks as used on turbocharged models.

should contact the fast idle stop, but should not depress over-ride spring. From this point, lengthen the control rod by 1½ turns of turnbuckle. This should permit full contact with fast idle stop screw on pump. The stop screws on pump limit engine speeds, but the length of control rod must permit full travel.

Move the hand throttle to the rear and adjust the stop screw and jam nut located under control panel so that hand lever is stopped by stop screw and does not damage pump linkage. Be sure that adjustment of hand lever stop nut does not change engine idle speed.

The starting fuel control linkage used on 4430 models prior to engine serial number 445570 and 4630 models prior to engine serial number 445948 must be correctly adjusted. Three different

types of starting fuel control linkages have been used on these models. The earliest type (used on 4430 models prior to engine serial number 390174 and 4630 models prior to engine serial number 383899) is shown in Fig. 117. The later type linkage (used on 4430 models engine serial number 390174-427053 and 4630 models engine serial number 383899-430385) is shown in Fig. 118. The latest type of linkage (used on 4430 models engine serial number 427054-445569 and 4630 models engine serial number 430386-445947) is shown in Fig. 119. All later turbocharged models have hydraulic aneroid activator integral with injection pump and do not use external mechanical starting fuel control linkage.

To adjust starting fuel control linkage for earliest type (4430 before

390174 and 4630 before 383899), refer to Fig. 117 and proceed as follows: Be sure that linkage is correctly adjusted, then move throttle to slow idle position. Loosen the set screw in stop collar, then pivot bellcrank rearward until starting fuel control shaft is moved to outward position. Pull starting fuel control rod toward rear, move stop collar forward against control rod swivel, then tighten set screw. Check adjustment by moving throttle to fast idle position and injection pump shut-off lever rearward (stop position). Move hand throttle lever back to idle position while checking the starting fuel control shaft at pump. As slow idle position is reached, the control shaft will move outward and a distinct click will be heard.

To adjust starting fuel control linkage for 4430 models (engine serial number 390174-427053) and 4630 models (engine serial number 383899-430385), refer to Fig. 118 and proceed as follows: Move the fuel shut-off control to the rear and check the linkage. The cable should rotate the bellcrank (1) to the latched position as shown. If control shaft (3) did not click into outer position as the bellcrank latched, loosen control bracket at aneroid, reposition bracket, then tighten mounting nuts. After bellcrank latching is correctly set, move the governor control lever (4) forward so clearance between lever and fast idle screw (5) is 0.100 inch. Loosen the set screw in collar (6), move collar against swivel of governor control lever, then tighten set screw. Move the governor control lever (4) until it contacts high speed stop screw (5) and check to be sure that bellcrank (1) is unlatched.

To adjust starting fuel control linkage for 4430 models (engine serial number 427054-445569) and 4630 models (engine serial number 430386-445947), refer to Fig. 119 and proceed as follows: End collar should be in-

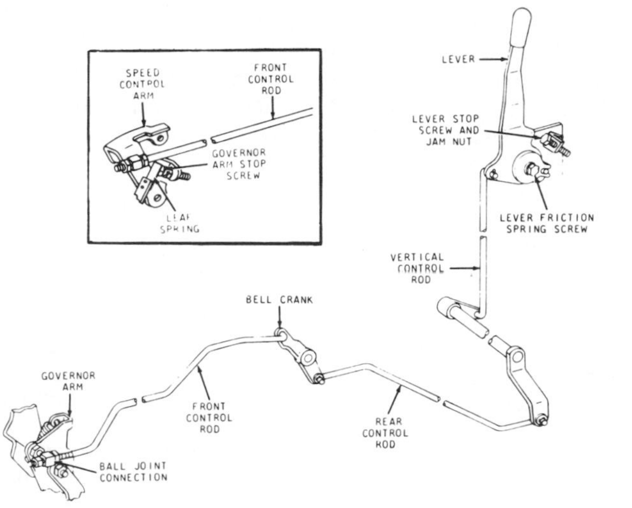

Fig. 115—View of Gasoline throttle linkage used on 4030 tractors. 4230 models have minor difference, which is shown in inset.

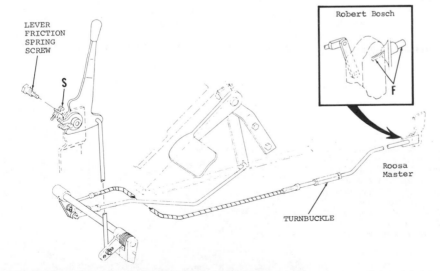

Fig. 116—View of typical throttle linkage used on diesel tractors. Optional foot control linkage and cable control used with "Sound-Gard" cab are shown. Fast idle stop for Robert Bosch pumps is shown at (F), idle stop for linkage is shown at (S). Idle speed stop on pump is not shown.

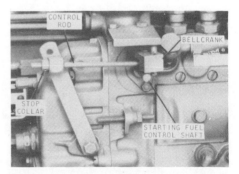

Fig. 117—View of injection pump starting fuel control linkage typical of type used on 4430 models before engine serial number 390174 and 4630 models before engine serial number 383899.

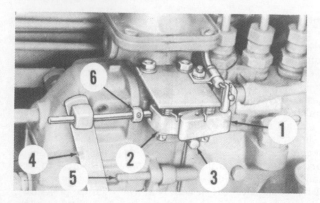

Fig. 118—View of injection pump starting fuel control linkage of type used on 4430 models engine serial number 390174-427053 and 4630 models engine serial number 383899-430385.

1. Bell crank
2. Latch
3. Control shaft
4. Governor lever
5. High speed stop screw
6. Collar

stalled on control rod so that rod protrudes 1/10-inch from rear of collar. Move fuel shut-off control to the rear (stop) position and check pump linkage. The bellcrank (1—Fig. 120) should be in latched position as shown. If bellcrank is not latched, move the rear cable clamp in slotted hole of guide bracket. If the control shaft did not move completely outward as the bellcrank latched, loosen the stud nuts and reposition the bracket further out away from aneroid, then tighten mounting nuts and recheck. Insert 0.090-0.110 inch thickness of feeler gage as shown at (G—Fig. 119) between governor control lever (4) and fast idle screw (5). Push governor control lever forward against gage (G), loosen set screw in front collar (6), slide collar (6) back against swivel and tighten set screw. Check adjustment by pulling shut-off control to stop position and checking latching. Push shut-off control in to run position, move speed control forward to fast idle position and check for unlatching. Readjust collar (6) if incorrect.

On 4430 models after engine serial number 445569, and 4630 models after engine serial number 445947, mechanical interlocking fuel control linkage is not used. Instead, these later models are equipped with hydraulic aneroid activator. Most repair and adjustment should be accomplished with pump removed from engine and attached to a test stand. Slow engine acceleration and excessive smoke when accelerating can be caused by incorrectly adjusted aneroid. Slow engine acceleration can also be caused by loose or broken inlet pipe or fitting to aneroid, cracked aneroid cover or defective aneroid diaphragm.

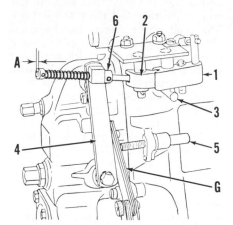

Fig. 119—View of injection pump starting fuel control linkage of type used on 4430 models engine serial number 427054-445569 and 4630 models engine serial number 430386-445947. Refer also to Fig. 120 for these models.

A. Protrusion (1/10 in.)
G. Feeler gage (1/10 in.)
1. Bell crank
2. Latch
3. Control shaft
4. Governor lever
5. High speed stop screw
6. Collar

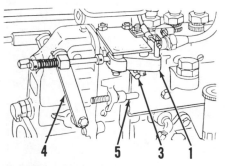

Fig. 120—View of linkage shown also in Fig. 119. Refer to text for adjustment.

1. Bell crank
3. Control shaft
4. Governor lever
5. High speed stop screw

126. SPEED ADJUSTMENT. Refer to Figs. 122, 123 and 124 for location of high speed adjustment screw (5) and low idle speed adjustment screw (8). Adjust the linkage as described in paragraph 125, then refer to the following:

On models with Roosa-Master pump, the low idle stop screw (8—Fig. 122 or Fig. 123) and the high speed stop screw (5) stop positions of the pump. The pump control lever (4) is equipped with a spring loaded override to prevent damage to pump while assuring full travel to contact stop screws. Low idle speed should be 780-820 rpm and high idle should be set to 2660 rpm for 4030 models, 2380-2420 for all other models with Roosa-Master pump.

Robert Bosch Series A-2000 pumps are used on 4430 models; Robert Bosch Series A-3000 pumps are used on 4630 models. There are several differences in these pumps, but the idle speed stop screw (8—Fig. 124) is located in similar location for all models. Slow idle speed should be approximately 780-820 rpm. To adjust idle speed, loosen locknut (11) and back screw (10) out about 3 or 4 turns. Loosen locknut (9), then turn slow idle stop screw (8) as required to set speed about 20 rpm less than desired slow idle rpm and tighten locknut (9). Turn spring loaded screw (10) in until idle rpm increases 20 rpm, then tighten locknut (11). Increasing the idle speed about 20 rpm with screw (10) should prevent slow speed surging. If surging continues at idle speed, stop engine, remove screw (10) and inspect spring for damage and freedom of movement in hollow of screw (10). Do not increase idle speed above 850 rpm in an attempt to smooth out idle.

The high speed stop screw (5) is located toward outside of pump on 4430 before engine serial number 445570 and 4630 before engine serial number 445948. The high speed stop screw is on

Fig. 122—Inset shows location of governor control lever (4), high speed stop screw (5), slow speed stop screw (8) and locknuts (7 & 9). Pump shown is for 4030 models. Refer to Fig. 123 for other 4230 models with Roosa Master.

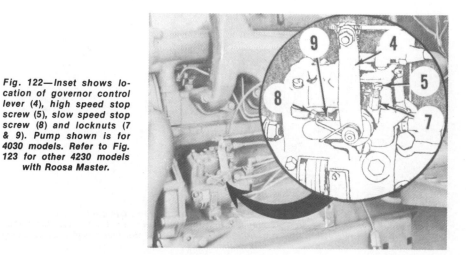

engine side of pump on late 4430 and late 4630 models. High idle stop screw should be set to limit speed to 2325-2425 rpm. Be sure that locknut (7) is tightened after speed is correctly set.

GASOLINE ENGINE GOVERNOR

The centrifugal, flyweight type gasoline engine governor is mounted on left side of engine and is driven from a timing gear idler on both the 4030 and 4230 models. Model 4030 uses a counterbalance arm and spring to prevent throttle lever from creeping.

4030 Model

128. LINKAGE ADJUSTMENT. Bring engine up to operating temperature. Be sure carburetor fast idle and slow idle adjustments are correct before any other adjustments are made. Move throttle lever full forward and make sure fast idle of approximately 2700 rpm is reached. Adjust fast idle stop screw if necessary. Stop engine,

disconnect speed change rod from speed change lever (11—Fig. 126) and move lever (7) all the way forward after disconnecting throttle rod (5) that comes from carburetor. Adjust both rods until the connections line up with holes in both levers, then shorten both rods by one turn and reconnect. This will make sure that both governor levers and carburetor will reach full throttle. Start engine and move throttle rearward until 800 rpm is reached and adjust slow idle stop screw against throttle lever if necessary.

129. REMOVE AND REINSTALL. To remove the governor assembly, first loosen clamps on inlet hose to carburetor, disconnect the hose and move air cleaner inlet pipe out of the way. Remove the two linkage rods and counterbalance arm spring, remove two retaining cap screws then lift off governor housing (1—Fig. 126) and associated parts. Weight shaft (15) and thrust sleeve (13) will usually remain with engine, but can be withdrawn.

When installing the governor assembly, position weight and gear assembly in tractor, install a new seal, then install housing and associated parts over weight unit. Tighten the retaining cap screws to a torque of 20-25 ft.-lbs. and adjust the governor linkage as outlined in paragraph 128.

130. OVERHAUL. Governor weights (14—Fig. 126) are a free fit on weight pins, but weight pins should fit tight in carrier. Examine rounded ends of weight feet which contact sleeve (13) for flat spots, and examine surface of sleeve (13) for wear or ridging. Renew weights if flat spots are evident, as this is the most common cause of governor failure. Weights are available individually, but should be renewed in pairs. Bushing (3) should be threaded for pulling and should be bottomed when installing. Press on closed end of bearing (2) or lettered end of bearing (4) when installing. Refer to paragraph 129 for governor installation procedure and to paragraph 128 for linkage adjustment.

4230 Model

131. LINKAGE ADJUSTMENT. Bring engine to operating temperature. Be sure carburetor slow idle adjustment of 800 rpm is correct before making any other adjustments. Move throttle lever full forward. Loosen two nuts on front control rod to governor and adjust fast idle to 2400 rpm. Move throttle rearward until 800 rpm is reached and make certain throttle shaft stop screw on carburetor is making contact with throttle body. While engine is at 800 rpm, adjust governor arm stop screw to give 1/32-inch clearance between screw and leaf spring. Adjust throttle lever slow idle stop

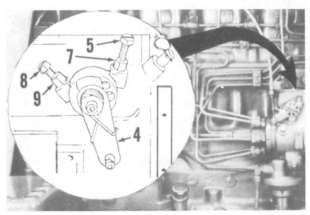

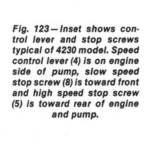

Fig. 123—Inset shows control lever and stop screws typical of 4230 model. Speed control lever (4) is on engine side of pump, slow speed stop screw (8) is toward front and high speed stop screw (5) is toward rear of engine and pump.

Fig. 124—View of injection pump controls and adjustment screws. Governor control lever (4) is located on engine side of pump on late 4430 and late 4630 models. Early models are similar except that high speed stop screw (5) is toward outside. Refer to text for adjustment.

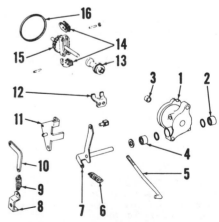

Fig. 126—Exploded view of gasoline engine governor and associated parts used on 4030 model.

1. Housing	9. Spring
2. Bearing	10. Counterbalance arm
3. Bushing	11. Speed change lever
4. Bearing	12. Fork
5. Throttle rod	13. Sleeve
6. Spring	14. Weights
7. Lever	15. Shaft and gear
8. Bracket	16. Seal

screw under access panel to contact throttle lever. Adjust throttle lever friction spring screw to where approximately 8 lbs. of force is required to move lever.

132. REMOVE AND REINSTALL. To remove the governor assembly, first loosen clamps on inlet hose to carburetor, disconnect the hose and move air cleaner inlet pipe out of the way. Remove the linkage rods to governor, remove three cap screws that hold governor to engine and lift off governor and associated parts.

When installing the governor, install a new seal and place governor assembly on engine. Use sealer on the three cap screw threads before installing. Adjust governor linkage as outlined in paragraph 131.

133. OVERHAUL. Refer to Fig. 127 for parts identification. Inspect all shafts, bushings and bearings for wear. Renew as necessary. Bearings (15) can be removed with JDE 11 puller if necessary. Governor drive shaft (11) O.D. should be 0.7495-0.7505 inch at bushing area. Bushing I.D. should be 0.7520-0.7540 inch; I.D. of thrust bearing (5) is 0.4500-0.4530 inch; Drive shaft (11) O.D. at thrust bearing is 0.4360-0.4380 inch; I.D. of weight carrier (7) is 0.4965-0.4985 inch. Shaft (11) O.D. at weight carrier is 0.4990-0.5010 inch. When reassembling, make sure that the loose end of thrust ball bearing (5) goes toward governor weight arms, and that lip of seal (14) is facing inward. After internal parts are assembled, place Woodruff key into drive shaft (11) and press drive gear on. End play of shaft and gear should be 0.006-0.012 inch. Refer to paragraph 132 for governor installation procedure and to paragraph 131 for linkage adjustment.

COOLING SYSTEM

RADIATOR

All Models

135. To remove radiator, drain cooling system and remove vertical air stack, muffler, side panels, grilles, screens and hood. Remove all brackets attached to radiator, disconnect hydraulic fluid line clamps and remove the screws retaining the fan shroud. Remove air cleaner hose if necessary, and disconnect radiator hoses. Remove radiator retaining cap screws and slide radiator out right or left side of tractor after fuel return line is removed. Install by reversing the removal procedure.

FAN AND WATER PUMP

All Models

136. REMOVE AND REINSTALL. To remove fan and/or water pump, drain cooling system and remove vertical air stack, muffler, hood, side panels and grilles. Remove the screws attaching fan shroud to radiator and cap screws attaching fan to pump hub; then, slide fan and shroud together out left side of tractor. Loosen fan belts, disconnect by-pass line if so equipped and lower radiator hose. Unbolt and remove the water pump. Install by reversing the removal procedure.

137. OVERHAUL. To disassemble the removed water pump unit, first remove hub (9—Fig. 128) using a suitable puller which attaches to two fan screw holes; then remove shaft, seal and impeller as an assembly from pump housing using a press. On 4030 models, hub and pulley (9 and 10) are one piece and must be pressed off together. Press shaft and bearing assembly out of impeller and remove ceramic sealing insert and rubber cup from impeller bore.

All parts are available individually except that ceramic sealing insert and cup for impeller is sold as a part of seal kit.

Coat bearing with light oil and press into housing bore until bearing is flush with front edge of housing. Coat outside of seal with Permatex of a type that is resistant to heat and ethylene-glycol anti-freeze and press seal into housing until it bottoms.

Apply a light coat of Permatex to shaft bore of impeller. Make sure seal lip and ceramic insert face are perfectly clean and apply a coat of light oil to insert. On 4030 models press impeller onto shaft until edge of impeller is flush with machined edge of impeller bore. On all other models, insert a feeler gage between impeller blades (4) and machined surface of housing (7). Clearance should be 0.015-0.025 inch.

NOTE: Support front end of shaft on bed of press when installing impeller; and impeller end of shaft when installing fan hub or pulley.

Install fan pulley and rear cover, then reinstall water pump and associated parts by reversing the removal procedure.

THERMOSTAT AND WATER MANIFOLD

All Models

138. The thermostats are contained in a thermostat housing in water manifold. Refer to Fig. 129 for view of typical manifold. All models except 4030 use double thermostats. On all models except 4230 to remove thermostats, drain system, detach upper radiator hose at front and remove

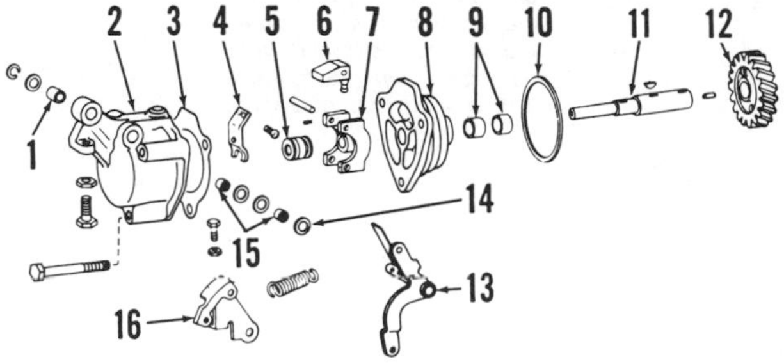

Fig. 127—Exploded view of 4230 model gasoline engine governor and associated parts.

1. Bushing
2. Housing
3. Gasket
4. Governor lever
5. Thrust ball bearing
6. Weight
7. Weight carrier
8. Drive shaft housing
9. Bushings
10. Seal
11. Drive shaft
12. Drive gear

13. Governor arm
14. Oil seal

15. Needle bearings
16. Speed control arm

Fig. 128—Exploded view of typical water pump. On 4030 models, hub and pulley are a one piece unit.

1. Gasket
2. Rear housing
3. Gasket
4. Impeller
5. Insert with cup
6. Seal
7. Pump housing
8. Shaft and bearings
9. Fan hub
10. Pulley
11. Fan blades
12. Belt

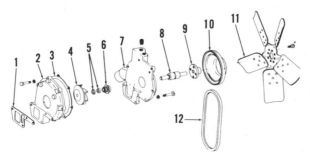

housing cap screws, then remove hose and housing as a unit. On 4230 models, remove 4 cap screws holding water outlet collector to thermostat housings, remove water by-pass pipe connectors from engine, and lay assembly to one side. If a thermostat is suspected of being faulty, check temperature range of unit suspected and test in heated water with a thermometer to be sure opening temperature is correct.

ELECTRICAL SYSTEM

ALTERNATOR AND REGULATOR

All Models

141. Delco-Remy "DELCOTRON" alternator is used on most models; however, a 90 amp John Deere alternator may be installed on some models. Delco-Remy 37 amp (1102359), 55 amp (1100491) and all of the 72 amp alternators used include an internally mounted solid state regulator. Voltage regulator for 90 amp John Deere alternator is attached to outside of alternator.

CAUTION: Because certain components of the alternator can be damaged by procedures that will not affect a D.C. generator, the following precautions MUST be observed.

a. When installing batteries or connecting a booster battery, the negative post of battery must be grounded.

b. Never short across any terminal of the alternator or regulator unless specifically recommended.

c. Do not attempt to polarize the alternator.

d. Disconnect all battery ground straps before removing or installing any electrical unit.

e. Do not operate alternator on an open circuit and be sure all leads are properly connected before starting engine.

STARTING MOTOR

All Models

142. Delco-Remy and John Deere starting motors are used on all models. Delco-Remy 1109147 starter is used on gasoline 4030 and 4230 models. Delco-Remy 1113391, 1113402, and 1113672 or John Deere AR55639 and AR77254 starters are used on diesel models.

DISTRIBUTOR

4030 & 4230 Gasoline Models

145. **TIMING.** The ignition is correctly timed when the "S" mark or correct degree mark on crankshaft pulley aligns with a mark on right side of timing gear cover at specified advanced timing rpm. Suggested method is to use a power timing light with engine running at specified rpm. Ignition

should occur at 24 crankshaft degrees BTC at 2500 engine rpm on 4030 gasoline models; at 20 crankshaft degrees BTC at 2000 rpm on 4230 gasoline models.

Recommended breaker point gap is 0.018-0.022 inch with cam angle 36-48 degrees for 4030 models. Recommended point gap is 0.020-0.024 inch with cam angle 22-26 degrees for 4230 models.

146. **REMOVE AND REINSTALL.** Before removing the distributor assembly, remove the distributor cap and turn engine crankshaft until front (No. 1) piston is at TDC on compression stroke. Disconnect distributor to coil primary lead wire, remove distributor clamp screw, then withdraw distributor assembly from cylinder block.

Proceed as follows to install distributor in 4030 gasoline models: Check to be sure that crankshaft is positioned at TDC on compression stroke of front (No. 1) cylinder. Position distributor housing so that primary lead is pointing directly toward engine, then turn distributor shaft so that rotor is pointing about 75 degrees clockwise from primary lead as shown in Fig. 132. Install the distributor into block bore fully making sure that gears mesh. When correctly installed, distributor rotor will point about 50 degrees clockwise from primary lead. An error of

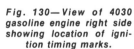

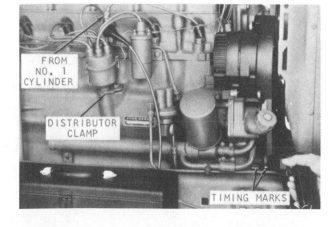

Fig. 130—View of 4030 gasoline engine right side showing location of ignition timing marks.

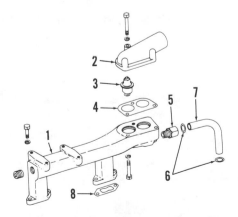

Fig. 131—View of 4230 gasoline engine right side showing location of ignition timing marks.

Fig. 129—Exploded view of typical water manifold. See text for differences between models.

1. Water manifold
2. Thermostat cover
3. Thermostat (2)
4. Gasket
5. Connector
6. "O" ring
7. Bypass pipe
8. Gasket (2)

one tooth will make about 30 degrees difference in rotor position. Firing order is 1-5-3-6-2-4.

Install distributor in 4230 gasoline models as follows: Check to be sure that crankshaft is set with front (No. 1) piston at TDC, then install distributor with offset drive tang aligned with coupling in cylinder block. Firing order is 1-5-3-6-2-4.

On all models, check running ignition timing as outlined in paragraph 145.

147. OVERHAUL. Refer to Fig. 133 for exploded view of distributor used

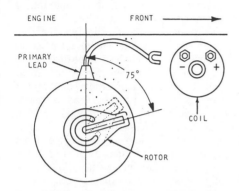

Fig. 132—Install distributor on 4030 gasoline models with front piston at TDC of compression stroke and rotor at 75 degrees from primary lead which is pointing toward engine block as shown. Because of the cut of distributor drive gear teeth, rotor will turn counter-clockwise to about 50 degree position when distributor is seated in block bore as indicated by dotted marks.

on 4030 models; or to Fig. 134 for exploded view of distributor used on 4230 models.

On 4030 models with Prestolite distributor, drive shaft bushings are renewable. New bushings should be pressed into housing until 0.094 inch below flush with housing. Drive shaft end play should not exceed limits of 0.002-0.010 inch. Centrifugal advance is changed by bending tangs which hold outer ends of advance springs. Light spring should control advance below 350 distributor rpm and heavy spring should be adjusted to change advance above 350 distributor rpm. Advance specifications are as follows in distributor rpm and advance angle: 0 degrees at 200 rpm; 6 degrees at 350 rpm; 15 degrees at 1200 rpm. Breaker point gap should be 0.018-0.022 inch with cam angle of 36-48 degrees for 4030 models.

On 4230 gasoline models, the Delco-Remy distributor is driven by an offset drive tang by the oil pump drive shaft. Distributor shaft diameter is 0.4895-0.4900 inch when new and new shaft should be installed if worn to less than 0.4875 inch. Housing bushings are not renewable separately, but housing should be renewed if bushing diameter exceeds 0.494 inch. Distributor shaft end play should be limited to 0.002-0.010 inch by adding or removing shims between drive coupling and bottom of distributor housing. New drive coupling

must be drilled for installation of spring pin. Coupling offset tang must be correctly positioned in relation to rotor position as shown in Fig. 135. Centrifugal advance in distributor degrees and distributor rpm should be as follows: 0-1 degree at 200 rpm; 1-3 degrees at 250 rpm; 5-7 degrees at 350 rpm; 14-16 degrees at 1200 rpm. Breaker point gap should be 0.020-0.024 inch with cam angle of 22-26 degrees for 4230 models.

148. SPARK PLUGS. Recommended spark plugs for 4030 gasoline models are Champion N11Y, Prestolite 14GT42 or AC45XL. Recommended spark plugs for 4230 gasoline models are Champion D-14, AC C-83, Prestolite 184. Electrode gap should be 0.025 inch for all models.

START-SAFETY SWITCH ADJUSTMENT

150. The start-safety switch is located high on right side of transmission housing, on all but Power Shift models, which has switch located in the transmission control valve.

To adjust switch, place transmission in neutral, remove switch and place enough washers under switch to close switch when reinstalled. Remove switch again, remove one washer and check selector in all positions. Switch should only close in the neutral and park positions. On Power Shift tractors it may be necessary to remove one additional washer. Switch will not operate properly on a Syncro Range or Quad Range tractor if shifter camshaft endplay exceeds 0.005 inch (Fig. 140).

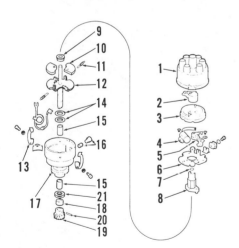

Fig. 133—Exploded view of Prestolite distributor used on 4030 gasoline models.

1. Cap	11. Advance springs
2. Rotor	12. Drive shaft
3. Cover	13. Clip
4. Condenser	14. Washers & spacers
5. Breaker points	15. Bushings
6. Base plate	16. Oil wick & plug
7. Retaining wire ring	17. Housing
8. Cam	18. Spacer
9. Spacer	19. Drive gear
10. Weights	20. Roll pin
	21. Washers

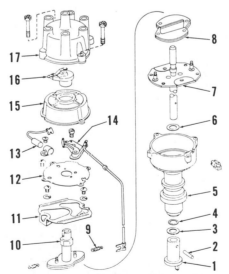

Fig. 134—Exploded view of Delco-Remy distributor used on 4230 models.

1. Drive coupling	9. Advance springs
2. Roll pin	10. Cam
3. Spacer washer	11. Hold down plate
4. Shim	12. Base plate
5. Housing	13. Condenser
6. Spacer washer	14. Breaker points
7. Drive shaft	15. Cover
8. Weights	16. Rotor
	17. Cap

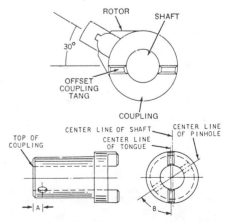

Fig. 135—New drive coupling for Delco-Remy distributor must be drilled as shown for installing roll pin shown at (2—Fig. 134).

Fig. 140—Start-safety switch adjustment should include end play check of shifter camshaft.

ENGINE CLUTCH (PERMA-CLUTCH)

NOTE: This section covers only the engine clutch, which is called a PERMA-CLUTCH and is used in tractors equipped with Syncro-Range, Quad-Range and Creeper transmissions. For models equipped with Power Shift transmission, refer to paragraphs 195 through 216.

The engine clutch includes two hydraulically applied power take-off clutch plates. When engine is not running, no hydraulic pressure is available, so clutch can not be applied and tractor can not be push-started.

Refer to the appropriate following paragraph for adjustment and to paragraph 167 for overhaul data.

LINKAGE ADJUSTMENT

All Models

160. TRANSMISSION CLUTCH. To adjust the transmission clutch linkage, remove left side cowl panel, and refer to Fig. 150. Correct distance (D) is 5¼ inches. Be sure pedal return spring is holding operating arm against its stop. If necessary to change adjustment, remove cotter pin and washer from clutch operating rod, and change position of the hex adapter in clutch pedal to obtain correct pedal height.

161. POWER TAKE-OFF. To adjust the power take-off clutch linkage, remove left side cowl panel and refer to Fig. 150. Pull pto control lever to within ¾-inch from rear of slot in instrument panel. Check pto operating rod to make sure that rear hole of three in pto lever is used. Rotate pto operating arm (on pressure regulating valve housing) clockwise until arm is against upper stop. Adjust lower yoke on rod until pin hole is lined up with operating arm hole then reinstall pin and tighten jam nut. On some tractors

adjustment is made at upper end of operating rod.

PERMA-CLUTCH PRESSURE TESTS

All Models So Equipped

In order for the Perma-Clutch to function properly and give long, trouble-free operation, it is very important that it be supplied with plenty of clean oil. Before making pressure checks to locate a malfunction, make certain that the filter is not plugged, and that the filter relief valve (Fig. 151) which is located on the left side of the tractor, just behind the filter, is free and operating properly.

162. CLUTCH PRESSURE CHECKS. To make a clutch pressure check, install a 300 lb. gage in plug hole marked "clutch" (Fig. 152). With tractor in PARK, engine at operating temperature and running at recommended rpm, oil pressure should be as follows:

Model	Test Rpm	Pressure (psi)
4030	1900	105-115
4230	1900	135-145
4430	1000	160-170
	1500	165-175
	2000	170-180
4630	1000	125-135
	1500	130-140
	2000	135-145

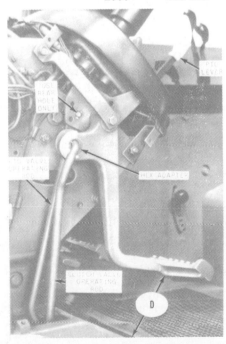

Fig. 150—View of clutch pedal and pto lever linkage adjustments on all models except Power Shift.

If pressure is below specifications, test both the pto brake and pto clutch at plugs shown. If pressure is normal at these points, there is leakage in the clutch valve circuits or 2-speed planetary circuit. If pressure is low at all three check points, increase and decrease engine speed approximately 300 rpm. If pressure rises and falls with engine speed, there is an internal leak or insufficient oil flow. If pressure changes only slightly with rpm change and all three test points show low pressure, adjust system pressure as follows: Remove pressure regulating valve plug (Fig. 152). Shims are available to adjust pressure, and each shim (19—Fig. 160) should boost pressure approximately 5 psi.

When pressure is satisfactory, depress clutch pedal while observing gage. Pressure should fall to zero when pedal is depressed, and come up gradually as pedal is slowly released, if clutch valve is working properly.

If pressure will not come up to

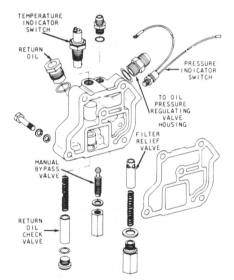

Fig. 151—Exploded view of filter relief valve housing. Check filter relief valve to make sure it is free.

Fig. 152—View of typical clutch regulating valve housing. Remove clutch plug and install gage to check Perma-Clutch pressure.

acceptable level, refer to transmission pump test, paragraph 275 in Hydraulic Section.

Refer to paragraph 259 for overhaul procedures.

163. PTO CLUTCH AND BRAKE. If pto is suspected of having a malfunction, refer to Fig. 152 and install a 300 psi gage in the pto clutch plug hole. Start engine and engage pto clutch. Pressure should be within limits that follow:

Model	RPM	Pressure (psi)
4030	1900	105-115
4230	1900	135-145
4430	2000	175-185
4630	2000	135-145

Engage pto brake and observe clutch pressure. Pto clutch pressure should drop to zero. When pto clutch is engaged gradually, pressure should rise gradually.

To test pto brake pressure, install another 300 psi gage in pto brake plug hole. With engine again running, apply the pto brake. Pressure should equal system pressure at specified rpm. Slowly push lever forward to the engaged position. Before clutch pressure starts to rise, brake pressure should drop to zero.

If pressure cannot be brought up to specifications, test transmission pump as outlined in paragraph 275 in Hydraulic Section. If pump tests good, remove pressure regulating valve housing (Fig. 152) and refer to Fig. 159A. Use an air hose with at least 125 psi pressure and a rubber tip on air gun to test each circuit as shown. A leaking circuit will be found when air can be heard escaping in any one passage. Move the 2-Speed planetary lever and apply air in both underdrive and direct

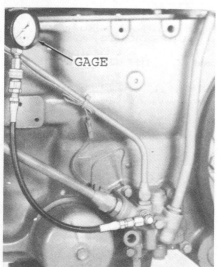

Fig. 152A—Attach gage to port as shown for checking lubrication pressure. Refer to text.

positions to check for leak in planetary circuit.

164. LUBRICATION PRESSURE AND FLOW. Models with Perma-Clutch use a transmission pump that flows a volume of approximately 10 gpm. On early models, lubrication pressure can be checked at either port marked "LUBE" as shown in Fig. 152 or at port where gage is shown attached in Fig. 152A. On late models, valve housing does not have lubrication pressure test port and pressure gage must be connected as shown in Fig. 152A. On early models, lubrication pressure should not be less than 15 psi when oil is at 150 degrees F. and engine is running at 1900 rpm. On later models, pressure should be checked with engine operating at 1500 rpm and with engine operating at 2000 rpm. Correct pressure is 25 psi at 1500 rpm with oil at 110 degrees F. or 17 psi with oil at 150 degrees F. Correct pressure at 2000 rpm is 43 psi at 110 degrees F. or 28 psi with oil at 150 degrees F.

Low lubrication oil pressure could be caused by: Pressure regulator to clutch valve housing gasket leaking; Adapter tube "O" rings leaking. If equipped with Quad-Range, additional possible causes are: Two speed shift valve leaking; Charge circuit malfunctioning.

The lubrication reduction valve can be checked with same gage attachment shown in Fig. 152A. Adjust engine speed until lubrication pressure indicated on gage is 10 psi, then depress clutch pedal and observe gage pressure. Proper operation of reduction valve is indicated by a dip in pressure. Check for stuck lubrication reduction valve if pressure does not dip.

165. OIL PRESSURE SWITCH. Remove oil pressure indicator switch (Fig. 151). Connect a hydraulic hand pump to switch and a 12 volt test light and battery across switch leads. The switch should open and close as follows:

Model	Open	Close
4030	83-93 psi	65-75 psi
4230	115-125 psi	90-100 psi
4430	145-155 psi	115-125 psi
4630	115-125 psi	90-100 psi

TRACTOR CLUTCH SPLIT

All Models

166. To detach (split) engine from clutch housing for access to engine clutch and flywheel, proceed as follows:

Drain cooling system and remove air stack, hood and muffler. Remove side shields, grille screens, cowl and step plates. Remove batteries, boxes and long connecting cable between batteries.

Discharge brake accumulator by opening the right brake bleeder screw and holding brake pedal down a few moments.

If tractor is equipped with air conditioning, disconnect the two couplers above oil filter on left side. Use two wrenches, one to hold coupler body and one to loosen coupler. If gas can be heard escaping, tighten coupler and loosen again. Keep coupler ends clean while disassembled.

On tractors with Perma-Clutch, remove the hex drive shaft for transmission pump to prevent bending shaft when tractor is separated. To remove hex shaft, remove three point hitch center link attaching bracket on rear of transmission housing. Remove the large plug which contains the rear bearing for the hex shaft and draw the long hex shaft rearward with pliers.

Disconnect throttle rod, wiring, hydraulic lines, heater hoses, tachometer cable and temperature indicator sending unit. Be sure to cap all disconnected hydraulic fittings to prevent dirt entry. If tractor has a Perma-Clutch, place a drain pan under clutch housing to catch the oil from clutch as tractor is separated. If tractor is equipped with front end weights, remove the weights. Support engine and transmission securely and separately, remove the connecting cap screws and roll transmission assembly rearward away from engine.

To attach, reverse the above procedure and tighten the connecting cap screws to a torque of 85 ft.-lbs. on ½-inch screws, or 300 ft.-lbs. on ¾-inch screws. Bleed steering system as outlined in paragraph 14 or 22.

R&R AND OVERHAUL PERMA-CLUTCH

All Models

Unless clutch has been slipping or giving some indication of trouble, DO NOT disassemble, as a properly adjusted and oil cooled assembly will give trouble free service for a very long time. Check the three transmission clutch levers for adequate clearance between the underside of levers and the clutch cover by driving wedges under the outer edge of levers as shown in Fig. 153. If some clearance does not exist with levers applied, unit should be overhauled. Transmission clutch levers will contact clutch cover before clutch discs are worn enough to damage other parts of the assembly. The clutch would then slip badly, and if transmission clutch levers were readjusted to compensate for the wear on the discs, serious damage could occur to the friction surfaces of clutch backing plate and separator plates with further wear.

167. Split tractor as outlined in paragraph 166. If clutch is to be overhauled, remove the six cover cap screws and disassemble unit. (If clutch is being removed to allow other work to be done, refer to the above procedures). Remove locknuts and adjusting nuts on the three clutch operating levers (Fig. 153). Refer to Fig. 154 and separate clutch pack, using care to avoid damaging discs and friction surfaces. Inspect all friction surfaces for scoring, cracking or sharp edges which could damage clutch discs. Smooth any raised area with crocus cloth. Original thickness of disc is 0.137-0.143 inch. Renew any disc that is not at least 0.110 inch in thickness, has less than 0.005 inch grooves in its face, or has chunks missing from friction surface. A dark color of disc lining is normal after use, and should not be considered sufficient reason to renew the disc, if thickness and groove depth are within specifications.

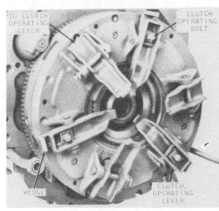

Fig. 153—Insert wedges under clutch operating levers as shown and check clearance under inside of lever.

When reassembling, place clutch pressure plate (6—Fig. 154) with friction surface up, insert the three clutch operating bolts (7) from the underside and place three blocks under bolts as shown in Fig. 155. Note that the transmission separator plate (3—Fig. 154) is thicker than pto separator plate (12). Install transmission clutch hub, disc, separator plate, second disc and backing plate in that order. Two separator plates (3) and three clutch discs (2) are used on 4630 models. Refer to Fig. 156 and install clutch pressure plate return springs, pto clutch hub, pto disc, pto separator plate, and second disc as shown. Make sure that the three tabs on pto clutch pressure plate are installed over the three return springs (Fig. 157). Install the clutch cover, three long clutch levers, bearing bars and adjusting nuts (Fig. 158). Tighten the three adjusting nuts enough to hold the clutch backing plate snugly against the clutch cover. This will squeeze the inside pto disc so that it will not fall off pto hub when clutch pack is installed in flywheel. When clutch pack is being installed, align the slots in separator plates with pins in flywheel. It may be necessary to slightly loosen the adjusting nuts in order to move separator plates. Install only four cap screws opposite each other in clutch cover.

168. **ADJUSTMENT.** It is very important that clutch adjustment be accurately performed, if uneven wear and hot spots are to be avoided. Since the Perma-Clutch is hydraulically applied, the clutch levers must fully release when apply pressure is removed from the apply piston. This will insure that the discs will not be overheated by a partially applied condition when engine is running with clutch released. The position of the clutch levers on the

clutch cover is very critical, as it determines the point at which the apply piston and bearing first make contact with levers. JDE-78 clutch adjusting tool with spacer plates is required for proper adjustment (Fig. 159). The spacer plate numbers to be used for each tractor model are listed in accompanying table.

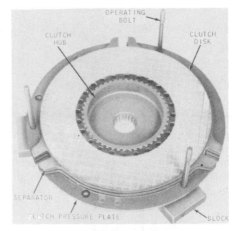

Fig. 155—Insert blocks under operating bolts as shown.

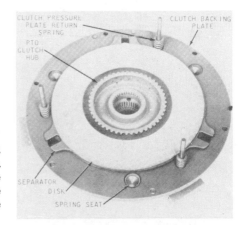

Fig. 156—Align separator plate slots with holes in backing plate as shown.

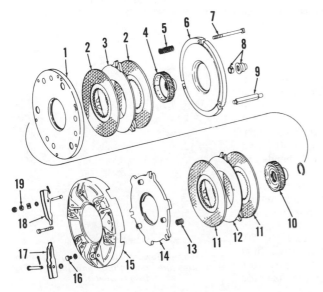

Fig. 154—Exploded view of Perma-Clutch and associated parts typical of all models. On 4630 models, two separators (3) and three discs (2) are used.

1. Backing Plate
2. Clutch disc
3. Separator, clutch
4. Clutch hub
5. Spring
6. Pressure plate
7. Operating bolt (3)
8. Pilot adapter
9. Flywheel pin (3)
10. Pto hub
11. Pto disc
12. Separator, pto
13. Spring
14. Pto pressure plate
15. Clutch cover
16. Adjusting screw
17. Pto clutch lever
18. Clutch lever
19. Adjusting nut

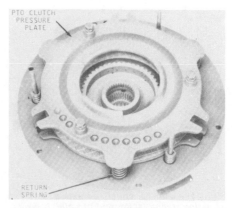

Fig. 157—Place pto pressure plate tabs on return springs as shown.

Model	Pto Clutch	Engine Clutch
4030, 4230	Gage 78-12 (0.070 in.) + Gage 78-3 (0.466 in.)	Gage 78-4 (0.253 in.) + Gage 78-10 (0.204 in.)
4430 (Prior to 47852)	Gage 78-12 (0.070 in.) + Gage 78-3 (0.466 in.)	Gage 78-4 (0.253 in.) + Gage 78-10 (0.204 in.)
4430 (After 47851)	Gage 78-12 (0.070 in.) + Gage 78-3 (0.466 in.)	Gage 78-4 (0.253 in.)
4630 (Prior to 18811)	Gage 78-12 (0.070 in.)	Gage 78-11 or 78-11A
4630 (After 18810)	Gage 78-12 (0.070 in.)	None Used

Make certain that only the proper plates for each tractor are used. Loosen the adjusting nuts on the three transmission clutch levers. Loosen three locknuts on pto levers and turn the adjusting screws in. Install the two mounting studs for adjusting tool in the two remaining clutch cover holes, which should be directly opposite from each other. Install adjusting tool so that the long screw will pass through the hole in cross bar and at the same time the three adjusting tool screws will bear on a flat part of pto pressure plate, and not on the lip. Leave the three adjusting tool screws loose and tighten the bar screw to 20 ft.-lbs. torque. Tighten the three adjusting tool screws evenly to 30 in.-lbs. torque, to load the pto pressure plate.

To adjust pto levers, make a mark or number on pto lever and pivot pin so that they will be returned to the same location that they were in when adjusted. Install one pto lever and pivot pin. Move lever out until it contacts outer adjusting surface, and turn adjusting screw out until it contacts the underside of lever. Carefully remove pivot pin and lever without moving adjusting screw and hold the screw from moving while locknut is tightened. Recheck setting by reinstalling pto lever and pin. If clearance is more than 0.010 inch, readjust lever to less than 0.010 inch. Pivot pin will go in tight because tightening the locknut will cause adjusting screw to stretch slightly. When all three levers have been adjusted using this procedure, loosen the three adjusting tool screws, but leave the bar screw tight.

Loosen the nuts on the three transmission clutch operating bolts and make sure the operating bolt heads are in their slots in pressure plate so that they will not turn. Move the transmission clutch levers out until they contact the **INNER** adjusting surface and finger tighten adjusting nuts. Tighten adjusting nuts in several steps (40, 70, then 90 in.-lbs. suggested) until final correct torque of 90 in.-lbs. is correct for all three nuts. Snug the locknuts

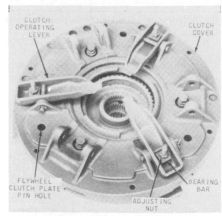

Fig. 158—Use a screwdriver through pin hole to align separator plates.

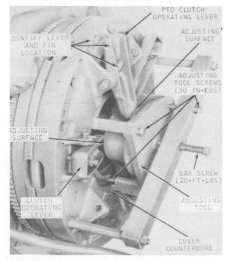

Fig. 159—JDE-78 clutch adjusting tool for adjusting transmission and pto clutch levers. Note two different adjusting surfaces for levers.

onto adjusting nuts, remove the adjusting tool, and carefully tighten locknuts while making sure that adjustment does not change. Install the two remaining clutch cover cap screws and tighten to a torque of 35 ft.-lbs.

Since the transmission clutch and pto clutch are applied by hydraulic pressure through needle thrust bearings, the bearings are running any time either clutch is applied. These needle bearings and their races should be carefully inspected for wear, scoring, or flattened rollers. Renew any bearing assembly that is not in perfect condition.

Rejoin tractor as outlined in paragraph 166, and bleed steering system as outlined in paragraph 14 or 22.

LUBRICATION REDUCTION VALVE

All Models With Perma-Clutch

When the transmission oil is cold and thick, the lube oil which is being routed to the transmission clutch discs tend to keep the discs rotating after clutch is released, making shifting difficult. The lubrication reduction valve (Fig. 162 or 163), located on the backside of clutch operating housing, prevents this condition by cutting off lube oil to the discs when clutch is released. When pto clutch is engaged with transmission clutch released, the valve is only partially opened, allowing a reduced volume of lube oil to transmission clutch, with most of the oil directed to pto discs.

170. **R&R AND OVERHAUL.** The early 4630 lubrication reduction valve and spring (not stop pin and piston) can be reached from the outside by removing the clutch regulating valve housing (2—Fig. 160), and reaching up inside

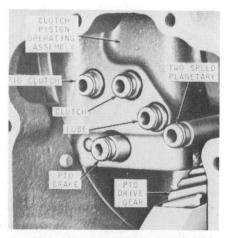

Fig. 159A—View of clutch and pto housing with pressure regulating housing (2—Fig. 160) removed.

Fig. 160—Exploded view of clutch regulating valve housing and associated parts. Flange gasket for item 2 not shown.

1. Plug
2. Outer housing
3. Shield
4. Clutch shaft
5. Pto shaft
6. Links
7. Springs
8. Pins
9. Springs
10. Clutch valve
11. Pto clutch valve
12. Valve housing
13. "O" rings
14. Adapters (4)
15. Adapter (Quad-Range)
16. Plug (Syncro-Range)
17. Plugs
18. Washers
19. Shim
20. Inner springs
21. Outer spring
22. Cooler relief valve
23. Clutch pressure valve
24. Gasket
25. Pto lock piston

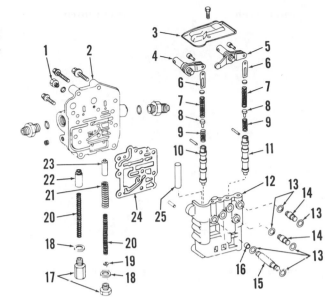

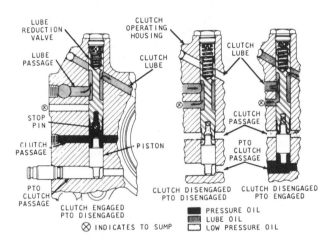

Fig. 162—Schematic view of lubrication reduction valve used on all models except early 4630. Refer to Fig. 163 for early 4630.

clutch housing. On all other models, split tractor as outlined in paragraph 166, which will allow access to clutch operating housing (Fig. 164). Remove six housing cap screws on outer rim of operating housing, and remove five inner cap screws, if equipped with Quad-Range transmission, or remove three inner cap screws if equipped with Creeper transmission. (Early 4630 models may have only four outer cap screws, and no inner cap screws to remove before operating housing is removed). Remove operating housing, and if only the lubrication reduction valve is to be removed, no further disassembly is needed. Remove the large hex plug at the rear of the clutch operating housing (Fig. 165). Remove spring and valve. The stop pin must be driven out before piston can be removed. Check valve and piston for wear or scoring and check spring for pitting or poor tension. When reassembling, drive stop pin into housing from the rear, until pin is flush with housing. Reassemble in reverse order of disassembly. Rejoin tractor as outlined in paragraph 166, and bleed steering system as outlined in paragraph 14 or 22.

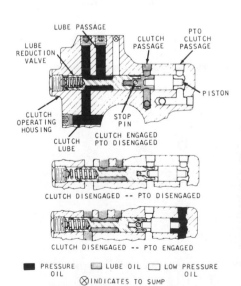

Fig. 163—Sectional view of lubrication reduction valve used on early 4630 models. Serial numbers 007000 and later used vertical type shown in Fig. 162.

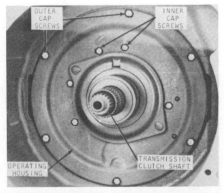

Fig. 164—View of clutch operating housing with Quad-Range transmission, showing inner and outer cap screws. Creeper transmission has only three inner cap screws.

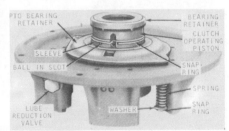

Fig. 165—View of removed clutch operating housing with front side up.

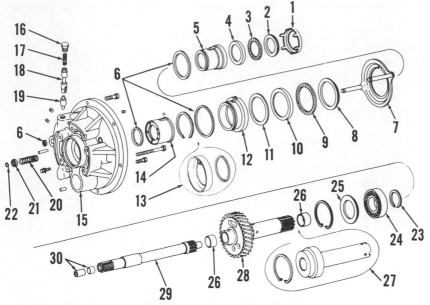

Fig. 166—Exploded view of clutch operating housing showing associated parts.

1. Bearing retainer	8. Outer race	16. Plug	23. Snap ring
2. Outer race	9. Needle bearing	17. Spring	24. Ball bearing
3. Needle bearing	10. Inner race	18. Lubrication	25. Oil shield
4. Inner race	11. Thrust washer	reduction valve	26. Bushing
5. Clutch operating	12. Pto operating piston	19. Piston	27. Adapter (no pto)
piston	13. Oil collector (no pto)	20. Spring	28. Pto drive shaft
6. Packing	14. Piston sleeve	21. Keeper	29. Clutch shaft
7. Pto bearing retainer	15. Operating housing	22. Retainer	30. Bushing (1 or 2)

CREEPER TRANSMISSION

Models So Equipped

The creeper transmission is a mechanically shifted planetary gearset, which gives an underdrive or direct gear ratio. The unit fits in front of the 8-speed transmission as an option on the Syncro-Range only. The optional creeper transmission is NOT available on 4630 model. In normal operations the unit is left in direct, but for operations where a very large speed reduction is needed, the unit can be shifted to underdrive. The underdrive may be used in the five lowest forward gears and both reverse gears. An in-terlock on the shift lever prevents the use of underdrive in forward gears 6, 7 and 8. The creeper shift lever must be in "Direct" in order to engage PARK position.

173. REMOVE AND REINSTALL. Split tractor as outlined in paragraph

166. Remove clutch operating housing (Fig. 164) as outlined in paragraph 170. Remove the small cap screw and retainer, which retains the planetary shifter shaft, from the right side of clutch housing. Pull shifter shaft out of housing far enough to clear the shifter assembly (Fig. 167). Pull planetary assembly and clutch shaft forward until assembly can be tilted away and removed for shifter assembly.

To reinstall, place planetary and transmission input shaft assembly (Fig. 167) into bearing quill and tilt sideways until shifter assembly can be installed with the shift yoke in REAR detent position. Use clutch shaft to guide entire assembly into splines of the transmission drive shaft. Move shifter assembly into the FRONT detent position and align assembly so that the hex shifter shaft can be pushed into the hex shaped hole in shifter arm. Shifter assembly should now be in proper position to accept three cap screws from clutch operating housing (Fig. 164). Install the cap screw and retainer which retains shift shaft in clutch housing and tighten to 20 ft.-lbs. torque. Reinstall clutch operating housing outer cap screws and be sure the three inner cap screws align with shifter assembly before final tightening.

Rejoin tractor as outlined in paragraph 166 and bleed steering system as outlined in paragraph 14 or 22.

174. OVERHAUL. Remove creeper planetary assembly as outlined in paragraph 173. Remove shifter collar (24—Fig. 168) from rear sun gear (20). Remove snap ring, collar plate (22), snap ring, rear sun gear, and transmission input shaft (25). Remove three

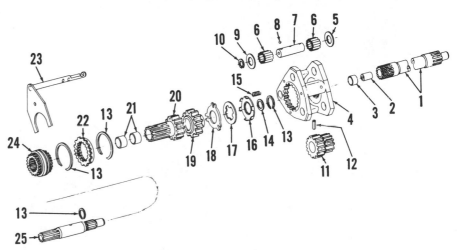

Fig. 168—Exploded view of Creeper planetary gears and associated parts.

1. Clutch shaft	8. Ball bearing (3)	14. Phenolic washer	20. Rear sun gear
2. Hex bushing	9. Washer	15. Spring (6)	21. Bushing (2)
3. Bushing	10. Snap ring (3)	16. Pressure plate	22. Collar plate
4. Planet carrier	11. Pinion gear (3)	17. Brake disc	23. Shifter yoke
5. Washer	12. Spring pin (3)	18. Backing plate	24. Shifter collar
6. Bearing (6)	13. Snap ring	19. Front sun gear	25. Transmission input
7. Pinion shaft (3)			shaft

Fig. 167—View of Creeper transmission planetary and shifter assembly after clutch operating housing is removed.

snap rings (10) from planet carrier (4) and remove three planet pinion shafts (7), being careful not to lose the three ball bearings (8). Remove pinions and bearings (11 and 6) and front sun gear (19). Drive out the three spring pins (12) far enough to remove snubber backing plate (18), brake disc (17), pressure plate (16) and springs (15). Inspect all parts for scoring and wear. Pressure plate (16) should be a minimum of 0.049 inch thick, and brake disc (17) should measure 0.075 to 0.085 inch. Springs should have a free length of approximately 1-1/16 inch, and should register 10.8 to 13.2 lbs. pressure when compressed to a length of 13/16-inch. Remove snap ring and clutch shaft (1) and inspect bushing (3) in shaft. If necessary to renew, press new bushing in 0.040 inch below the end of shaft.

Reassemble in reverse order of disassembly and refer to Fig. 168 as guide. First install hex bushing (2) into clutch shaft (1), install clutch shaft into carrier (4) and install snap ring to retain shaft. Install thrust washer (14) on clutch shaft and place carrier in a vise with clutch shaft pointed down, or in a hole in a bench so that unit will be held steady. Assemble the springs, pressure plate, disc brake and backing plate. Use a press (Fig. 169) to compress the

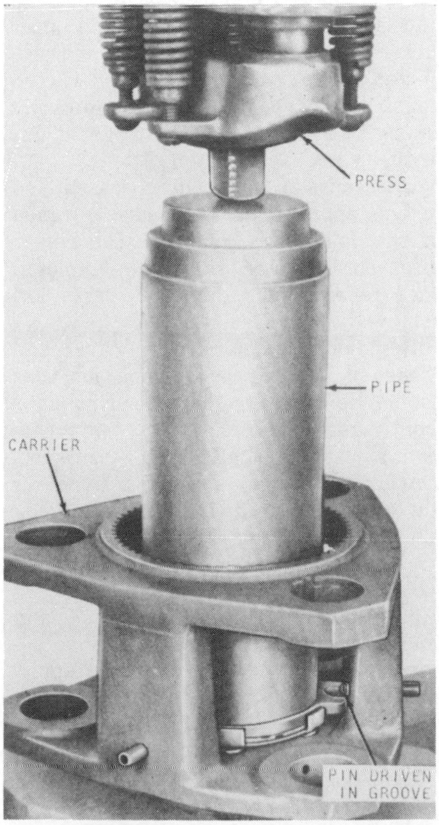

pieces until the three spring pins can be driven into planet carrier (4—Fig. 168). About 13/32-inch of pin should stick into the inside of carrier to retain backing plate. Install front sun gear (19) into carrier so that the notches align with tangs on disc brake. Refer to Fig. 170 and install planet pinions so that the timing marks on all pinions align with the three marks on sun gear. This will allow the output sun gear teeth to mesh with pinions. Install snap ring on transmission input shaft and insert forward splines into front sun

gear. If output (rear) pinion bushings must be renewed, press new bushings (21—Fig. 168) to 0.055 inch below sun gear end of pinion, and 0.037 inch below splined end and install gear. Install large snap ring into inner groove in carrier, install collar plate (22) and the other snap ring into outer groove. Place shifter collar (24) onto splines of output sun gear and install assembly as outlined in paragraph 173.

SYNCRO-RANGE (8-SPEED GEAR) TRANSMISSION

(For Quad-Range Models, Refer to Paragraph 192)

(For Creeper Transmission, Refer to Paragraph 173)

(For Power Shift Models, Refer to Paragraph 195)

The "Syncro-Range" transmission is a mechanically engaged transmission consisting of three transmission shafts and a single, mechanically connected, remote mounted control lever as shown in Fig. 179. The four basic gear speeds are selected by coupling one of the differential bevel pinion shaft idler gears to the splined bevel pinion shaft, and on all models, can only be accomplished by disengaging the engine clutch and bringing the tractor to a stop. The high and low speed ranges within the four basic speeds are selected by shifting only the synchro-

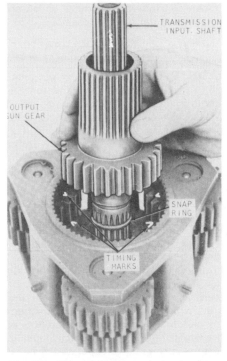

Fig. 170—View of timing marks on front sun gear and pinion gears. Marks must align as shown so that rear sun gear teeth will mesh properly

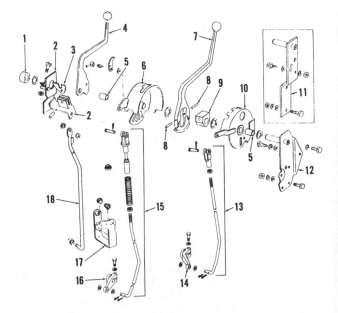

Fig. 169—Compress snubber brake springs with a press as shown, to install spring pins to retain backing plate.

Fig. 179—Syncro-Range and Creeper transmission shift levers and associated parts. Items 4, 17 and 18 are used only if equipped with Creeper transmission.

1. Spacer
2. Support
3. Shift latch
4. Creeper shift lever
5. Bushing (2)
6. Speed change quadrant
7. Shift lever
8. Spring pin (2)
9. Pivot
10. Speed range quadrant
11. Support
12. Support
13. Speed range rod
14. Speed range arm
15. Speed change rod
16. Speed change arm
17. Bell crank
18. Creeper shift rod

nized coupler on the transmission drive shaft and, because of the design, can be accomplished by disengaging the engine clutch and moving the control lever, without bringing tractor to a halt.

NOTE: The rotating speeds of transmission drive shaft and idler gears are automatically equalized by the synchronizing clutches. All other phases of shifting are under the direct control of the operator. The fact that clashing of gears is eliminated by the synchronizing clutches does not relieve operator of the responsibility of using care and judgment in re-engaging the engine clutch after gears have been shifted.

The idler gears and bearings on the main shaft and bevel pinion shaft are pressure lubricated by a separate transmission oil pump which also serves as the charging pump for main hydraulic system pump.

INSPECTION

All Models

180. To inspect the transmission gears, shafts and shifters, first drain the transmission and hydraulic fluid and remove operator's platform. Disconnect the control support brackets, Perma-Clutch control valve operating rod linkage and Quad-Range control valve linkage, then remove transmission top cover. Examine shaft gears for worn or broken teeth and the shift linkage and cam slots for wear.

CONTROL QUADRANT

This section covers disassembly and overhaul of shifter controls mounted on shift console to the right of operator. Removal, inspection and overhaul of shift mechanism inside the transmission housing is included with transmission gears and shafts.

All Models

181. **R&R AND OVERHAUL.** To overhaul the control quadrant, remove the console sheet metal by removing the six cap screws holding the vertical panel, and three cap screws under the right fender. Disconnect the shifter rods (13 and 15—Fig. 179), and Front Drive Rotary Switch rods on models so equipped; then unbolt and remove quadrant assembly.

If lever (7) or pivot (9) must be renewed, proceed as follows: Clamp the lower curved portion of lever in a protected vise and slip a 5/32-inch cotter pin inside the two spring pins (8). Refer to Fig. 180. Grasp spring pin with clamping pliers and extract with a

twisting motion. When reassembling, leave at least 7/16-inch of roll pin protruding from lever.

TRANSMISSION DISASSEMBLY AND REASSEMBLY

Paragraphs 182 and 183 outline the general procedure for removal and installation of the main transmission components. Disassembly, inspection and overhaul of the removed assemblies is covered in overhaul section beginning with paragraph 184, which also outlines those adjustment procedures which are not an integral part of assembly.

All Models

182. **DISASSEMBLY.** Any disassembly of transmission gears and controls requires that tractor first be separated (split) between engine and clutch housing as outlined in paragraph 166. Disconnect Power Front Wheel Drive drain pipe, if so equipped. Remove the transmission top cover, and all necessary wiring and hydraulic lines.

NOTE: Tractor may be equipped with either a two gear pto drive train system, or a four gear system. A four gear system tractor may be separated between clutch housing and transmission housing without first removing pto lower drive gear and shaft assembly. Be sure to remove the cap screws or hex nuts which must be reached from front of clutch housing and through top of transmission to rear of clutch housing, before attempting to separate the housings.

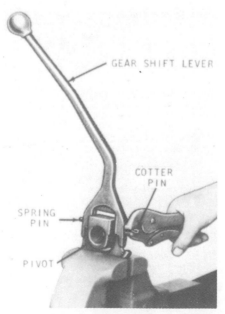

Fig. 180—Insert a cotter pin inside spring pin to aid in removal. Refer to paragraph 181.

A two gear system has a large lower pto drive gear inside clutch housing which, together with pto drive shaft, must be removed BEFORE housings are separated. Remove the rear bearing quill and pto shaft as an assembly if equipped with a 1000 rpm only pto. If unit is a dual speed, remove rear bearing quill, stub shaft and 540 rpm drive gear, which is directly under bearing quill. Compress large snap ring which holds pto drive shaft bearing in transmission housing and pull drive shaft rearward out of housing. The large pto drive gear in clutch housing will remain on a sleeve in the front drive shaft bearing. Slide the large gear backward off the sleeve, being careful not to damage the oil shield which is made onto the gear, then tip gear forward and remove toward the front. If oil shield is damaged and will rub on gear, separate the welds around shield halves and renew either half. Weld around edges by tacking halves in five places. To separate clutch housing, remove transmission top cover and the two top cap screws in backside of clutch housing. Remove remaining cap screws, support the rear of tractor under front of transmission housing, lift clutch housing with a chain hoist and roll rear of tractor away. Supports should be placed under front and rear of transmission housing so that it is very stable. Front drive shaft bearing and sleeve should be inspected after clutch housing is removed, and renewed at this time if necessary. Clutch housing may be attached temporarily to engine if it is necessary to get it out of the way.

Transmission must be disassembled in the approximate sequence outlined in the following paragraphs; however disassembly need not be completed once defective or damaged parts are removed.

Remove transmission top cover and rockshaft housing or transmission rear cover. Remove the detent spring caps from right side of housing, using caution, as caps are under spring pressure. Lower the upper shifter arm to its lowest position and remove special lock nut from inner end of shaft. Be careful not to drop parts in housing, as they will be difficult to remove. Refer to Fig. 181. Withdraw shifter arm and shaft from housing. Oil seal may be renewed at this time. Remove speed range shifters and rail. (4630 model has two rails).

Jack up one rear wheel of tractor and turn bevel ring gear until one of flat surfaces of differential housing is toward transmission oil pump, then unbolt and remove the manifold, oil tubes and pump.

Tape or clip the synchronizer clutches together to keep them from separating while input shaft is being removed. Working through input shaft front bearing retainer and using a brass drift, drive the input shaft (drive shaft) rearward to force rear bearing cup part of the way out of housing. Remove front bearing retainer using care to keep shim pack together and undamaged, then lift transmission input shaft assembly out top opening.

Remove lower (speed change) shifter cam and forks Fig. 182. Remove differential assembly as outlined in paragraph 222. If tractor is equipped with four gear pto drive train, remove retainer plate on pto lower idler gear and use a puller to remove gear and roller

bearing (front). The rear bearing cone will remain on differential pinion shaft. If tractor has the two gear pto drive train, remove retainer plate on pinion shaft. Before driving pinion shaft rearward, install a "C" clamp through hole in transmission housing (Fig. 183) and tighten against front (fourth and seventh speed) gear while using a brass drift or soft hammer to drive on shaft. As shaft begins to move, slide the parts forward, retighten "C" clamp and remove snap rings (Fig. 184) as they become exposed.

Continue to move shaft components and snap rings forward until rear snap ring has been unseated; then bump differential drive shaft (pinion shaft) rearward, lifting gears and associated parts out top opening as shaft is removed.

NOTE: The snap rings are the same diameter but of different thickness; the thickest snap ring being in rear groove. Thus no snap ring can fall into another groove as shaft is removed or installed.

Remove countershaft front bearing retainer (9—Fig. 189) and shim pack (8). Remove snap ring retaining rear bearing cup (1); then using a brass drift, drive the countershaft rearward until rear bearing cup is removed.

Countershaft can now be lifted out top opening of transmission case.

Overhaul transmission main components as outlined in paragraphs 184 through 191; assemble as outlined in paragraph 183.

183. **ASSEMBLY.** To assemble the transmission unit, proceed as follows: Install countershaft, rear bearing cup and retaining snap ring, then install front bearing retainer using the removed shim pack. Tighten retaining cap screws to a torque of 35 ft.-lbs.; then using a dial indicator, check countershaft end play. Adjust end play if necessary, to the recommended 0.001-0.004 inch by varying the thickness of shim pack (8—Fig. 189). Shims are available in thicknesses of 0.006, 0.010 and 0.019 inch for Model 4630; or 0.006, 0.010 and 0.018 inch for other models.

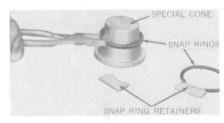

Fig. 185—Special assembly tools are needed to expand and install snap rings.

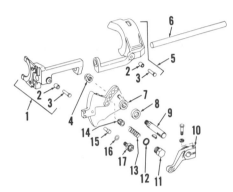

Fig. 181—Exploded view of Speed Range shifters and parts as used on 4230 and 4430 Syncro-Range and Quad-Range models. Other models are similar, except that 4630 has two rails (item 6).

1. Reverse shifter	10. Shift arm
2. Roller	11. Shifter pawl cap
3. Pin	12. Aluminum washer
4. Special nut	13. Spring
5. Low-High shifter	14. Shifter pawl
6. Shift rail	15. Safety switch
7. Shifter cam	follower
8. Oil seal	16. Shim
9. Shifter shaft & key	17. Start-safety switch

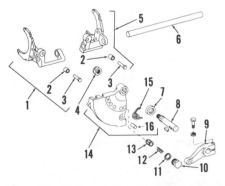

Fig. 182—Exploded view of Speed Change shifters and parts as used on 4230 and 4430 Syncro-Range and Quad-Range models. Other models are similar.

1. Rear collar shifter	9. Shift arm
2. Roller	10. Shifter pawl cap
3. Pin	11. Aluminum washer
4. Special nut	12. Spring
5. Front collar shifter	13. Shifter pawl
6. Rail	14. Shifter cam
7. Oil seal	15. Parking lock spring
8. Shifter shaft & key	16. Spring pin

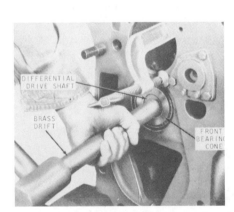

Fig. 183—Install a "C" clamp to hold front gear forward while front bearing is driven off shaft.

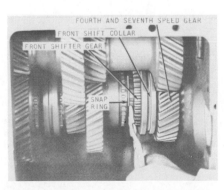

Fig. 184—Remove four snap rings as they become exposed as shaft moves rearward. Keep snap rings in order, with thickest to the rear.

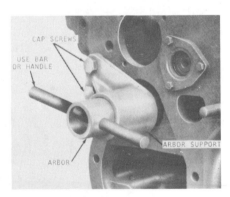

Fig. 186—Install arbor and support to hold parts while assembling differential pinion shaft gears.

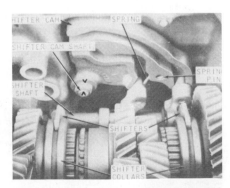

Fig. 187—Speed change shifter mechanism showing component parts. Note "V" alignment marks on shifter cam and shaft. Install shifter cam spring as shown.

If bevel pinion shaft or transmission housing are being renewed; or if shaft bearings require renewal or adjustment; refer to paragraph 191 for adjustment procedure.

Special tools are required to reinstall bevel pinion shaft. The needed tools are 1—JDT-1, snap ring pliers; 4—JDT-3A, expander plates, and 1—JDT-2 expander cone for Model 4030; or 1—JDT-1, snap ring pliers; 4—JDT-3 expander plates and 1—JDT-2 expander cone for other models. Expand the snap rings and install expander plates as shown in Fig. 185, then lay out snap rings in order according to thickness. Refer to Fig. 192 for order of reassembly of shaft components. Install arbor and support as shown in Fig. 186 (or other suitable support) to hold parts while assembling. Insert bevel pinion shaft (1—Fig. 192) through rear bore, install largest gear (5) and thickest snap ring (6) with expander plate in place. Continue to move shaft forward, installing remaining parts in proper order. Second thickest snap ring is installed next, and so on, with the thinnest at front. After fourth speed gear (14) is installed, remove expander plates and seat snap rings in their grooves; then install remainder of parts using the removed (or previously determined) shim pack (16). Tighten cap screws retaining end plate (20) to a torque of 35 ft.-lbs. on grade 5 cap screws, or 50 ft.-lbs. on grade 8 cap screws.

Install speed change shifter forks and cam, and insert shift rail through forks. Install camshaft and arm, making sure index marks are aligned as shown in Fig. 187. Tighten locknut securely then adjust camshaft end play to 0.002-0.007 inch if necessary, by loosening clamp nut and sliding actuating arm on shaft. Too much end play in shifter camshaft can allow detent springs to force the shift cam out of proper operating position. If shift cam spring (Fig. 187) has been removed or renewed, make sure the long end of spring is installed toward the shifter cam shaft.

To install transmission drive shaft and shifter mechanism assembly, place shaft in transmission housing and install front bearing cup, shims and retainer. Tighten cap screws securely, then install rear bearing cup and transmission oil pump. Tighten pump cap screws to 20 ft.-lbs. torque. Check transmission drive shaft end play using a dial indicator and adjust to 0.004-0.006 inch by means of shims (21—Fig. 190). Shims are available in thicknesses listed below.

4030 . . . . 0.006, 0.010 and 0.018 in.
4230 & 4430 . . . . 0.006 and 0.010 in.
4630 . . . . 0.006, 0.011 and 0.018 in.

Tighten front bearing retainer cap screws to 35 ft.-lbs. torque.

Reinstall speed range shifters and associated parts, making sure index marks are aligned and camshaft end play is within the recommended range of 0.002-0.007 inch.

OVERHAUL

The following paragraphs cover overhaul procedure of transmission main components after transmission has been disassembled as outlined in paragraph 182.

184. SHIFTER CAMS AND FORKS. Refer to Fig. 181 for an exploded view of speed range shifter mechanism and Fig. 182 for speed change shifters. Examine shifting grooves in cams (7—Fig. 181 and 14—Fig. 182) for wear or other damage. Parking lock spring (15—Fig. 182) is retained to cam by

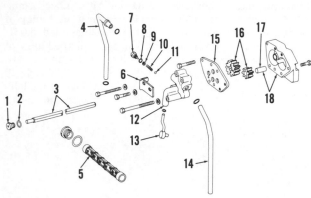

Fig. 188—Exploded view of typical Syncro-Range transmission oil pump and associated parts.

1. Plug w/bushing
2. "O" ring
3. Hex drive shaft
4. Outlet tube
5. Intake screen
6. Clamp
7. Plug
8. "O" ring
9. Shim
10. Spring
11. Ball
12. Manifold
13. Tube
14. Intake tube
15. Cover
16. Gears
17. Dowel pin
18. Body w/pin

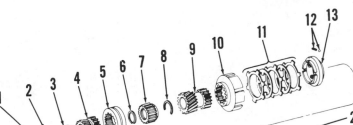

Fig. 189—Exploded view of transmission countershaft and associated parts.

1. Bearing cup
2. Bearing cone
3. Snap ring
4. Gear
5. Countershaft
6. Bearing cone
7. Bearing cup
8. Shim
9. Bearing retainer

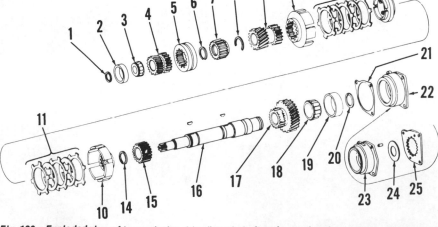

Fig. 190 — Exploded view of transmission drive (input) shaft and associated parts. Items 23, 24 and 25 are used instead of item 22 on models equipment with Creeper Transmission. On later models, low and high range drums (10) are not interchangable. High range drum is marked with groove. Internal plates (11) are splined for splined blocker (13) on later models.

1. Snap ring
2. Bearing cup
3. Bearing cone
4. Reverse range pinion
5. Shift collar
6. Snap ring
7. Reverse range drive collar
8. Snap ring
9. Low range pinion
10. High-Low range drum
11. Synchronizer plates
12. Spring and Ball
13. High-Low range blocker
14. Snap ring
15. High-Low range drive collar
16. Drive shaft
17. High range pinion
18. Bearing cone
19. Bearing cup
20. Snap ring
21. Shim
22. Bearing housing
23. Bearing housing
24. Thrust washer
25. Shifter collar plate

spring pin (16). The spring must have sufficient tension to shift the front shift coupling into engagement.

185. TRANSMISSION PUMP. The removed transmission pump may be disassembled by removing cover (15—Fig. 188). Check all surfaces, bushings and bushing bores. Lubricating check valve spring (10) should test 3.2-3.8 lbs. pressure at ¾-inch length. Clearance between driven gear and bushing should be 0.0015-0.0045 inch and cover should be flat within 0.003 inch. All parts are available individually.

186. COUNTERSHAFT. Refer to Fig. 189 for an exploded view of countershaft and bearings. The shaft is a one-piece unit except for high-speed gear (4). Gear is keyed to shaft and retained by snap ring (3). Countershaft should have 0.001-0.004 inch end play in bearings when properly installed.

187. TRANSMISSION DRIVE SHAFT. To disassemble the removed transmission drive shaft, remove snap ring (1—Fig. 190); then remove rear bearing cone (3) using a suitable press. Lift off reverse range pinion (4) and reverse range shift collar (5).

CAUTION: The four detent balls and springs (12) will be released when blocker is withdrawn. Use care not to lose these parts.

Remove snap ring (6) and use a press or puller to remove drive collar (7). Remove snap ring (8) and withdraw low range pinion (9); then remove high-

low synchronizer observing the precautions outlined in note above. High-low range drive collar (15) can be pressed from shaft after removing snap ring (14). High range pinion (17) can be removed after removing high-low range collar (15) or bearing cone (18).

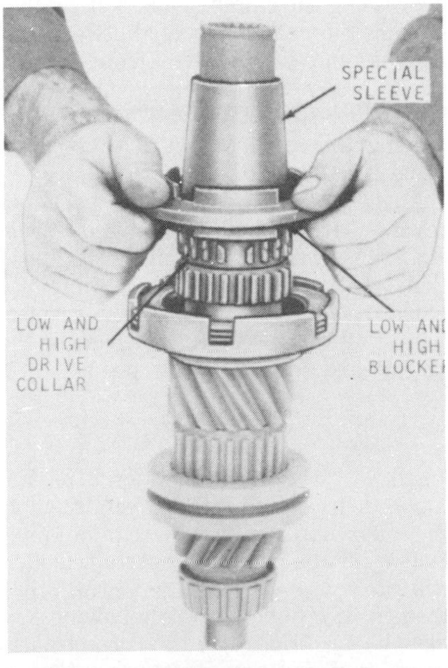

188. SYNCHRONIZER CLUTCHES. The purpose of the synchronizer clutches is to equalize the speeds of the transmission drive shaft and the selected range pinion for easy shifting without stopping the tractor. The synchronizer clutches operate as follows:

The range drive collars (7 or 15—Fig. 190) are keyed to the shaft. Synchronizer clutch drums (10) are splined to the range pinions. The blocker ring (13) is centered in drive collar slots by the detent assemblies (12). Synchronizer clutch discs (11) are connected alternately (by drive tangs) to the clutch drum and blocker ring. When engine clutch is disengaged and control lever moved to change gear speeds, the first movement of shifter linkage applies contact pressure to clutch discs (11) causing blocker to try to rotate on drive collar. Drive lugs inside the blocker ring ride up the ramps in drive collar preventing further movement of blocker ring until shaft and gear speeds are equalized. When rotative speeds are equal, thrust force on blocker ring is relieved and synchronizer drum couples gear to shaft without clashing.

189. INSPECTION AND ASSEMBLY. Inspect transmission drive shaft for scoring or wear in areas of range pinion rotation and make sure oil passages are open and clean. Inspect drive lugs of blocker rings and friction faces of blocker rings and synchronizer drums. Check synchronizer discs for wear using a micrometer. Renew the entire set if any disc measures 0.060

inch or less. Check drive tangs on disc for thickening due to peening, and renew disc if badly peened.

Reassemble the transmission drive shaft by reversing the disassembly procedure. A special installing cone JD-4A is required to install detent assemblies in blocker rings and blocker assemblies on drive collar; refer to Fig. 191.

190. BEVEL PINION SHAFT. Except for bearing cups in housing and rear bearing cone on shaft, the bevel pinion unit is disassembled during removal. Refer to Fig. 192 for exploded view.

All gears should have a diametral clearance of 0.004-0.006 inch on shaft. First-third gear contains a bushing but bushing is not available for service. The bevel pinion shaft is available only as a matched set with the bevel ring gear. Refer to paragraph 225 for information on renewal of ring gear. Refer to paragraph 191 for mesh position adjustment procedure if bevel gears and/or housing are renewed.

191. PINION SHAFT ADJUSTMENT. The cone point (mesh position) of main drive bevel gear and pinion is adjustable by means of shims (4—Fig.

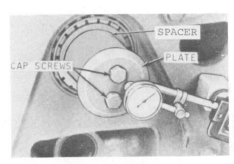

Fig. 193—Use a pipe spacer behind retainer plate to check preload with dial indicator. Refer to paragraph 191.

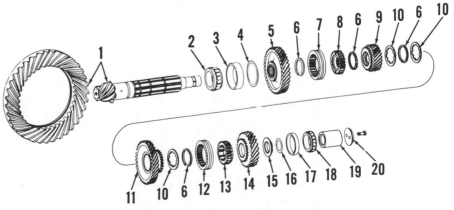

Fig. 192—Exploded view of bevel pinion shaft and associated parts.

1. Pinion shaft & gear	6. Snap ring (4)	11. 2nd - 5th gear	16. Shim
2. Bearing cone	7. Shift collar	12. Shift collar	17. Bearing cup
3. Bearing cup	8. Rear shifter gear	13. Front shifter gear	18. Bearing cone
4. Shim	9. 6th - 8th gear	14. 4th - 7th gear	19. Spacer (no pto)
5. 1st - 3rd gear	10. Thrust washer (3)	15. Thrust washer	20. Retainer plate

Fig. 191—Assembling the Low-High range synchronizer using special tool.

192) which are available in thicknesses of 0.003, 0.005 and 0.010 inch in all models except 4030, which uses 0.003 and 0.005 inch thicknesses only. The cone point will only need to be checked if transmission housing or ring gear and pinion assembly are renewed.

The correct cone point of housing and pinion are factory determined. To make the adjustment, refer to appropriate model below.

4030 MODEL. Add the standard guide number 1.755 to the number etched on pinion gear and subtract the result from 8.912. The difference will be the correct shim pack thickness.

4230 AND 4430 MODELS. Subtract the number etched on end of pinion gear from 8.326. The difference will be the recommended thickness (in inches) of shim pack to be installed.

4630 MODEL. Subtract the number etched on end of pinion gear from 9.5393. The difference will be the recommended thickness (in inches) of shim pack to be installed.

On all models use a punch and drive out rear bearing cup (3—Fig. 192) and add or remove shims from shim pack (4) until correct total thickness is obtained.

The bevel pinion bearings are adjusted to a preload of 0.004-0.006 inch by means of shims (16). If adjustment is required, it should be made before installing the gears as follows:

First make sure cone point is correctly adjusted as outlined and that cone (2) is bottomed on pinion shaft shoulder. Install the shaft, washer (15), the removed shim pack (16) plus one 0.010 inch shim and bearing cone (18). If tractor has a four gear pto drive train, install a suitable pipe spacer instead of the lower pto idler gear and bearings, install retainer plate (20) and the retaining cap screws. Measure shaft end play using a dial indicator. When disassembling, remove shims equal to the measured end play plus 0.005 inch. Assemble shaft and gears as outlined in paragraph 183.

QUAD-RANGE

Several production changes have occurred which effect service and, where known, these differences will be described. It is especially important to make sure that only correct parts are installed which are compatible with the tractor serial number. Also, be sure that parts for the correct model are installed, because very similar, but different, parts are sometimes used in other 4000 series tractors.

OPERATION

192. The "Quad-Range" transmission consists of an 8-speed mechanically engaged transmission which is similar to the "Syncro-Range" transmission; however, an additional 2-speed hydraulically shifted planetary unit is installed ahead of the 8-speed transmission.

The two hand levers control speed selection at one of three locations.

The lever furthest to the right, marked A, B, C, D and P moves gears on the bevel pinion shaft to engage one of four reduction ratios or "PARK". This shift CHANGE selector lever should ONLY be moved when the tractor is stopped and the clutch pedal is depressed.

The lever located between the throttle and the shift change selector lever is the RANGE selector lever. Front to rear movement of the range selector lever moves the shift couplings located on the transmission drive shaft to provide two forward reduction ratios and one reverse. Do not attempt to change direction of tractor by moving range shift lever between 1st and Reverse until tractor is stopped and clutch pedal is depressed. The range selector lever can be moved between 2nd and 3rd with tractor moving, but be sure to depress clutch pedal.

The RANGE selector also moves from side to side as it moves from front to rear. This lateral (left to right) movement moves a hydraulic control valve located in the "Quad-Range" housing. When the lever is to the left, hydraulic pressure is used to engage a clutch which results in direct drive. When the lever is to the right, hydraulic pressure engages a brake which causes the planetary gear assembly to drive the tractor at a reduced speed. Lateral movement alone can be accomplished with the tractor moving, and without depressing the clutch pedal. Notice, that moving the range selector between 2nd and 3rd moves the lever front to rear and laterally. Shifting between 2nd and 3rd may be accomplished with tractor moving; however, be sure that clutch pedal is depressed.

A mechanical interlock prevents reverse from being engaged when change selector lever is in the "D" position.

The accompanying table shows position of all three sections for each of the 16 forward and 6 reverse speeds.

TESTING

192A. Several production changes have occurred and some effect assembly procedures in such a way that assembly that would be correct on certain models will cause serious malfunction on other models. If trouble occurs following service, be sure that only the correct parts were used and that parts were assembled correctly for the specific model be-

SPEED SELECTION	QUAD-RANGE PLANETARY		SYNCRO-RANGE TRANSMISSION			CHANGE TRANS.			
	UNDERDRIVE	DIRECT	LOW	HIGH	REVERSE	A	B	C	D
A-1	X		X			X			
A-2		X	X			X			
A-3	X			X		X			
A-4		X		X		X			
B-1	X		X				X		
B-2		X	X				X		
B-3	X			X			X		
B-4		X		X			X		
C-1	X		X					X	
C-2		X	X					X	
C-3	X			X				X	
C-4		X		X				X	
D-1	X		X						X
D-2		X	X						X
D-3	X			X					X
D-4		X		X					X
A-1R	X				X	X			
A-2R		X			X	X			
B-1R	X				X		X		
B-2R		X			X		X		
C-1R	X				X			X	
C-2R		X			X			X	

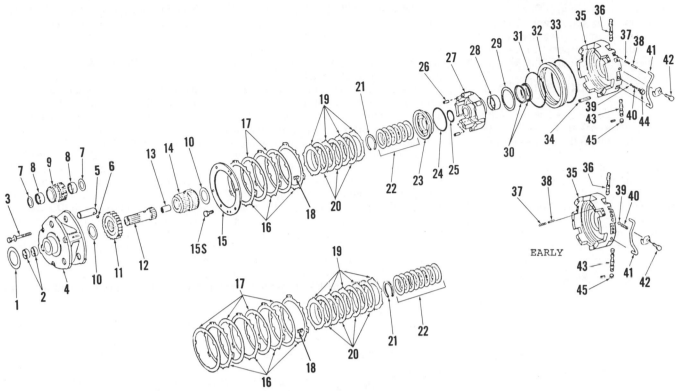

Fig. 193A — Exploded view of Quad-Range planetary, clutch and brake assembly. The brake housing (35) on early models is not only different parts, but is also assembled differently.

1. Thrust washer
2. Bushings
3. Screws
4. Planet carrier
5. Shaft
6. Steel balls (5 mm)
7. Thrust washers
8. Needle bearings
9. Planet pinion
10. Thrust washers
11. Sun gear
12. Input shaft
13. Transmission oil pump drive shaft bushing
14. Clutch shaft sun gear
15. Brake backing plate
16. Disc with facing (number required varies by model & serial number)
17. Separator plates (number required varies by model & serial number)
18. Springs (6 used)
19. Separator plates (number required varies by model & serial number)
20. Disc with facing (number required varies by model & serial number)
21. Snap ring
22. Belleville spring washers
23. Clutch piston
24. Sealing ring
25. Packing
26. Dowel pins (3 used)
27. Clutch drum
28. Bushing
29. Thrust washer
30. Sealing rings
31. Inner packing ring
32. Brake piston
33. Outer packing ring
34. Spring pins (6 used, 5/16 x 1¾ in)
35. Brake housing
36. Control valve
37. Detent spring
38. Detent plunger (for valve 36)
39. Detent ball (for valve 43)
40. Detent spring
41. Oil tube
42. Cap screw
43. Shift valve
44. Plug (late models)
45. Plug

ing serviced. One example of incorrect assembly that can cause problems is with the detent ball (39 – Fig. 193B and Fig. 193C). If the ball and spring for late models is installed under screw (42 – Fig. 193B) as it is on early models, the passage will be blocked and correct operation is impossible.

If problems occur after an extended time of proper operation, first check condition of oil filter and make sure that filter relief valve (Fig. 151) is free and operating properly. Without adequate volume and pressure, the planetary cannot operate properly.

To check system pressure, refer to Fig. 152 and install a 300 psi pressure gage in the plug hole marked "CLUTCH", which is located in the clutch regulating valve housing on left side of tractor. With the engine up to operating temperature and shift lever in **PARK** position, disconnect the planetary shift rod at bellcrank. Run engine at 1900 rpm and check the system pressure in both direct and underdrive by shifting the rod back and forth by hand and observing pressure gage. Indicated pressure should be as follows:

4030	105-115 psi
4230	134-145 psi
4430	170-180 psi
4630	135-145 psi

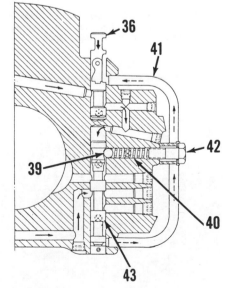

Fig. 193B — Cross-section showing location of detent ball (39) and related parts for early models. Refer to Fig. 193A for legend.

If pressure check reading is 10 psi or lower in direct than in underdrive, the direct clutch circuit is leaking. If underdrive pressure reading is 10 psi or lower than direct, the underdrive circuit is leaking. If both direct and underdrive pressures are low, the "O" rings on the 2-speed planetary adapter tube (Fig.

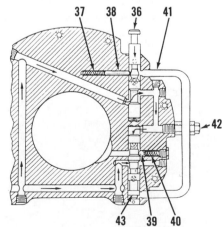

Fig. 193C — Cross-section showing location of detent ball (39) and plunger (38) on late models. Refer to Fig. 193A for legend.

159A) could be leaking. Check pressure as outlined in paragraph 162 and if necessary, remove pressure regulating valve housing and perform air tests as outlined in paragraph 163.

REMOVE AND REINSTALL

193. To remove the 2-speed planetary unit, tractor must be split between engine and clutch housing as outlined in paragraph 166. Remove the hydraulic lines and operating rods from pressure regulating valve housing. Remove housing and extract all five of the oil passage adapters (Fig. 159A). Refer to Fig. 164 and remove the six outer cap screws. Remove five inner cap screws that hold the 2-speed planetary to the clutch operating housing, and remove housing. If clutch shaft did not come out with housing, pull it out of 2-speed unit. On all early models except 4630, refer to Fig. 194 and remove pin in upper corner which retains shifter arm shaft. The shifter arm on late 4030, 4230 and 4430 models and all 4630 models is held to shaft by a spring pin which must be driven out before shaft and arm can be removed.

On early 4030, 4230 and 4430 models, the large pto gear must be removed before planetary pack will come out. First remove the cap screws and nuts which hold the pto rear bearing quill and remove quill. If equipped with the 1000 rpm only pto, remove the quill and pto shaft as an assembly. On dual speed units, remove the stub shaft, quill and bearing, then remove the 540 rpm gear, compress snap ring behind pto shaft gear and draw snap ring, bearing and pto shaft from tractor. The large pto gear and oil shield will stay on the sleeve in clutch housing. Push pto gear rearward off the sleeve, tip it forward and lift from housing, being careful not to damage oil shield. The planetary assembly, including the short transmis-

sion drive shaft with hex bushing can be pulled straight forward and removed.

The planetary assembly can be removed without removing pto drive gear and shield on late 4030, 4230, 4430 models and all 4630 models. Remove the hex bushing from planetary unit if it did not come out with the clutch shaft. Make a tool to pull the transmission input shaft, using a piece of rod about 36 inches long. Flatten one end of rod and bend end over just enough to form a hook, but small enough to go through shaft. Insert rod, hook the rear of input shaft and pull forward. Move and rotate splines so that shaft will come out easily. Planetary assembly can now be removed by pulling forward and rotating enough to clear pto gear.

Reinstall planetary assembly in reverse order of disassembly. Before installing assembly, make sure the rear thrust washer is installed. Insert hex bushing in the front of the short transmission input shaft and install shaft and bushing after the planetary assembly has been installed. Rotate gears to align

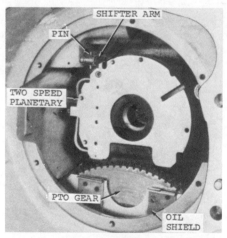

Fig. 194A — View of late Quad-Range 2-speed planetary clutch operating housing removed. Refer to Fig. 194 for early type.

splines if necessary. Complete assembly, and rejoin tractor as outlined in paragraph 166.

OVERHAUL

194. Remove 2-speed planetary unit as outlined in paragraph 193. Place assembly face down on bench and remove cap screws (3–Fig. 195) from planetary pinion carrier. Lift pinion carrier off dowel pins. After carrier is removed, the planet gears (9–Fig. 196) can be removed by pushing planet gear shafts (5) out toward front and catching the retainer balls (6) as shaft starts out of carrier. The sun gear (14) can be removed after planet gears are out of carrier. Inspect bearings, washers and shafts for excessive wear or pitting, and renew as necessary. Remove six cap screws which hold backing plate (15–Fig. 195) to underdrive brake housing. Remove piston return springs (18–Fig. 197 or Fig. 197A), center hub and sun gear (14), brake discs (17), separator plates (16), clutch discs (20), plates (19) and direct drive clutch drum (27). Note the thrust washer (29–Fig. 198) that fits between clutch drum and brake housing. Remove brake piston (32) with its inner and outer seals. The clutch piston spring washers (22–Fig. 199) must be compressed using JDT-24 compression tool or other suitable means,

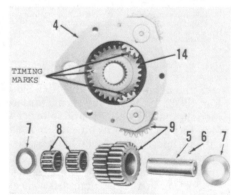

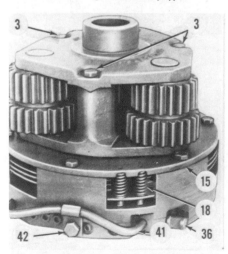

Fig. 195 — Remove three cap screws (3) to separate planet gear assembly. Early type is shown. Refer to Fig. 193A for legend.

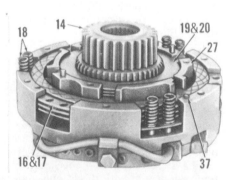

Fig. 197 — View of early model brake housing with backing plate removed. Note detent spring (37) location for early type housing. Refer to Fig. 197A for later type.

Fig. 194 — View of typical early Quad-Range 2-speed planetary with clutch operating housing removed. Refer to Fig. 194A for later models.

before snap ring (21) can be removed. Disassemble the shift valve and control valve using Fig. 193A, Fig. 200 and Fig. 200A as a guide. Air pressure may be used to blow out the retaining plug (45) after retaining pin has been driven out. Use care to avoid losing the detent spring (40) and ball (39).

On early models, detent ball and

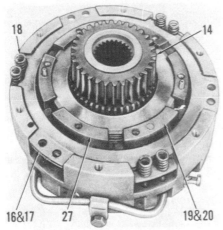

Fig. 197A — View of late model brake housing with backing plate removed. Discs and plates are installed alternately at (16 & 17) and (19 & 20).

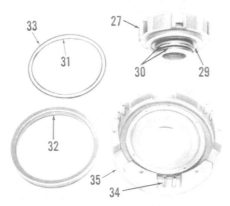

Fig. 198 — Brake housing (35) with direct drive clutch drum (27) and brake piston (32) removed. Note thrust washer (29) and pins (34).

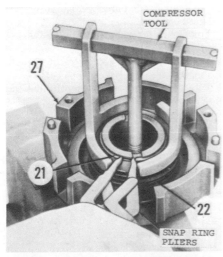

Fig. 199 — Compress Belleville spring washers (22) with suitable tool to remove snap ring (21).

spring for valve spool (43) are located under cap screw (42–Fig. 200) which clamps tube (41) in place. On later models, a special plug (44–Fig. 200A) retains the detent ball (39) and spring (40) in a separate bore. During original production this change in detent location occurred at approximately tractor serial number 016417 for 4030 model, 039806 for 4230, 071584 for 4430 and 030082 for 4630 models. The valve spools (36 and 43–Fig. 200 and Fig. 200A) are also different to accept the different location of the detent ball. If the complete planetary is exchanged, the later type may be installed in certain tractors before the listed serial number.

Before tractor serial number 015106 for 4030 models, 034664 for 4230, 060942 for 4430 and 025361 for 4630 models, the detent ball (39) is 5/16-inch diameter. Detent ball for later models is ¼-inch diameter.

Spring (37–Fig. 200 and Fig. 200A) should have free length of 1-5/16 inches for 4030 models after serial number 015105, for 4230 models after serial number 034663, for 4430 models after serial number 060941 and for 4630 models after serial number 025360.

Spring (40) should have free length of 1⅞ inches for 4030 models after serial number 016417, for 4230 models after serial number 033323 and for 4430

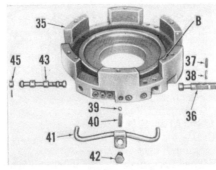

Fig. 200 — View of early model with valves and detent assemblies removed. Ball and spring (39 & 40) are located below plug (42) and plunger and spring (37 & 38) enter bore (B). Refer to Fig. 200A for later type.

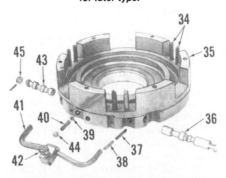

Fig. 200A — View of late model with valves and detent assemblies removed. Ball and spring (39 & 40) are located below special plug (44). Plunger and spring (37 & 38) enter bore for tube (41) and are located across bore for valve (36) as shown in cross-section Fig. 193C.

models after serial number 071584. The spring used on 4630 models is different.

Inspect control valve, shift valve and their bores. Any nicks or scores must be smoothed on valves or in bores to insure free movment of valves. Inspect all parts for wear, and pay particular attention to the needle bearings and their retainers. Bushings should be renewed if not in excellent condition. The minimum free height of each Belleville spring washer (22–Fig. 193A) should be 0.126 in.

To reassemble planet gears, install sun gear in carrier, making sure the timing "V" marks point toward the three pinion shaft bores (Fig. 196). Install the three planet gears, with timing marks aligned with marks on sun gear. Thrust washer must be in place. Install the balls in pinion shafts and recheck timing marks. Install the packing ring in clutch drum (Fig. 201) and sealing ring on clutch piston.

Install piston in drum and assemble spring washers facing each other as shown, so that the inner diameter of the end washers will contact the piston and snap ring. Be sure that correct number of spring washers is installed. Refer to the accompanying table for number originally installed.

Model & Serial No. of Spring Washers
No. Range   (22–Fig. 193A)
4030–
    Before S.N. 011989 8
    After S.N. 011988 6
4230–
    Before S.N. 026634 8
    After S.N. 026633 6
4430–
    Before S.N. 041801 8
    After S.N. 041800 6
4630–
    Before S.N. 015372 8
    015372 to 025360 6
    After 025360 8

Compress springs with compression tool (Fig. 199) and install snap ring. Lubricate all parts before assembling. Install brake piston seals and use JDT-21-1 outer sleeve (Fig. 202) and JDT-21-2 inner sleeve (all models except 4630) and install brake piston. 4630 models use JDT-23-1 and JDT-23-2 sleeves to install piston.

Install shift valve (43–Fig. 200 or Fig. 200A), retainer (45) and retaining pin. On late models (Fig. 200A), insert 1-5/16 in. long spring (37) and plunger (38) through bore for tube (41) and across bore for control valve. Compress spring by pressing on plunger (38) with a small rod, then insert control valve (36) into bore with flats on valve toward detent plunger. Insert detent ball (39) and spring (40) into third hole from bottom and install plug (44).

On early models, insert detent ball

(39–Fig. 200) and spring (40) into hole for screw (42).

On all models, install tube (41–Fig. 200 or Fig. 200A), clamp and screw (42). Tighten screw (42) to 21 ft.-lbs. torque.

When installing clutch drum in brake housing, be sure the sealing rings and thrust washer are in place (Fig. 198). Note that brake housing has three slots with spring pins, and three slots without pins (Fig. 202). Tangs on first separator plate (16–Fig. 193A) installed must be installed on pins to provide a seat for piston return springs (18–Fig. 197 or Fig. 197A). Install one brake disc (17–Fig. 193A), then install remaining separator plates (16) and discs (17) alternately. Tangs on remaining separator plates go into the slots without pins.

4030 (Before S.N. 011989)
Brake separator plates (16)–
 Number used 4
 Thickness, new . . . . 0.085-0.095 in.
Brake discs (17)–
 Number used 4
 Thickness, new . . . . 0.087-0.093 in.
 Minimum facing groove depth0.010 in.
Brake separator springs (18)–
 Free length . . . . 1.16 in.
 Working load . . . . 24-30 lbs. at 0.81 in.
Clutch separator plates (19)–
 Number used 6
 Thickness, new . . . . 0.055-0.065 in.
Clutch discs (20)–
 Number used 5
 Thickness, new . . . . 0.060-0.063 in.
 Minimum facing groove depth0.006 in.
Spring washers (22)–
 Number used 8

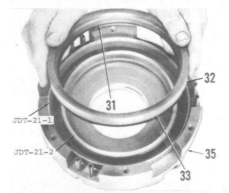

Fig. 201—Exploded view of clutch drum and associated parts. Assemble spring washers (22) as shown, so end washers will contact piston (23) and snap ring (21) with the inner diameter.

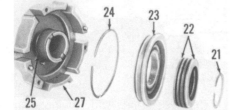

Fig. 202—Use the proper sleeve identification number when installing brake piston. See text for model application.

4030 (After S.N. 011988)
Brake separator plates (16)–
 Number used 3
 Thickness, new . . . . 0.085-0.095 in.
Brake discs (17)–
 Number used 2
 Thickness, new . . . . 0.087-0.093 in.
 Minimum facing groove depth0.010 in.
Brake separator springs (18)–
 Free length . . . . 0.71 in.
 Working load . . . . 24-30 lbs. at 0.52 in.
Clutch separator plates (19)–
 Number used 4
 Thickness, new . . . . 0.055-0.065 in.
Clutch discs (20)–
 Number used 3
 Thickness, new . . . . 0.071-0.075 in.
 Minimum facing groove depth0.006 in.
Spring washers (22)–
 Number used 6
 Minimum free height . . . . 0.126 in.

4230 (Before S.N. 026634)
Brake separator plates (16)–
 Number used 4
 Thickness, new . . . . 0.085-0.095 in.
Brake discs (17)–
 Number used 4
 Thickness, new . . . . 0.087-0.093 in.
 Minimum facing groove depth0.010 in.
Brake separator springs (18)–
 Free length . . . . 1.16 in.
 Working load . . . . 24-30 lbs. at 0.81 in.
Clutch separator plates (19)–
 Number used 6
 Thickness, new . . . . 0.055-0.065 in.
Clutch discs (20)–
 Number used 5
 Thickness, new . . . . 0.060-0.063 in.
 Minimum facing groove depth0.006 in.
Spring washers (22)–
 Number used 8
 Minimum free height . . . . 0.126 in.

4230 (After S.N. 026633)
Brake separator plates (16)–
 Number used 3
 Thickness, new . . . . 0.085-0.095 in.
Brake discs (17)–
 Number used 2
 Thickness, new . . . . 0.087-0.093 in.
 Minimum facing groove depth0.010 in.
Brake separator springs (18)–
 Free length . . . . 0.71 in.
 Working load . . . . 24-30 lbs. at 0.52 in.
Clutch separator plates (19)–
 Number used 4
 Thickness, new . . . . 0.055-0.065 in.
Clutch discs (20)–
 Number used 3
 Thickness, new . . . . 0.071-0.075 in.
 Minimum facing groove depth0.006 in.
Spring washers (22)–
 Number used 6
 Minimum free height . . . . 0.126 in.

4430 (Before S.N. 041801)
Brake separator plates (16)–
 Number used 4
 Thickness, new . . . . 0.085-0.095 in.
Brake discs (17)–

Number used 4
Thickness, new . . . . 0.087-0.093 in.
Minimum facing groove depth0.010 in.
Brake separator springs (18)–
 Free length . . . . 1.16 in.
 Working load . . . . 24-30 lbs. at 0.81 in.
Clutch separator plates (19)–
 Number used 6
 Thickness, new . . . . 0.055-0.065 in.
Clutch discs (20)–
 Number used 5
 Thickness, new . . . . 0.060-0.063 in.
 Minimum facing groove depth0.006 in.
Spring washers (22)–
 Number used 8
 Minimum free height . . . . 0.126 in.

4430 (After S.N. 041800)
Brake separator plates (16)–
 Number used 3
 Thickness, new . . . . 0.085-0.095 in.
Brake discs (17)–
 Number used 2
 Thickness, new . . . . 0.087-0.093 in.
 Minimum facing groove depth0.010 in.
Brake separator springs (18)–
 Free length . . . . 0.71 in.
 Working load . . . . 24-30 lbs. at 0.52 in.
Clutch separator plates (19)–
 Number used 4
 Thickness, new . . . . 0.055-0.065 in.
Clutch discs (20)–
 Number used 3
 Thickness, new . . . . 0.071-0.075 in.
 Minimum facing groove depth0.006 in.
Spring washers (22)–
 Number used 6
 Minimum free height . . . . 0.126 in.

4630 (Before S.N. 015372)
Brake separator plates (16)–
 Number used 5
 Thickness, new . . . . 0.085-0.095 in.
Brake discs (17)–
 Number used 5
 Thickness, new . . . . 0.087-0.093 in.
 Minimum facing groove depth0.010 in.
Brake separator springs (18)–
 Free length . . . . 1.42 in.
 Working load . . . . 24-30 lbs. at 0.99 in.
Clutch separator plates (19)–
 Number used 7
 Thickness, new . . . . 0.055-0.065 in.
Clutch discs (20)–
 Number used 6
 Thickness, new . . . . 0.071-0.075 in.
 Minimum facing groove depth0.006 in.
Spring washers (22)–
 Number used 8
 Minimum free height . . . . 0.126 in.

4630 (After S.N. 015371)
Brake separator plates (16)–
 Number used 4
 Thickness, new . . . . 0.085-0.095 in.
Brake discs (17)–
 Number used 3
 Thickness, new . . . . 0.087-0.093 in.
 Minimum facing groove depth0.010 in.
Brake separator springs (18)–
 Free length . . . . 0.96 in.
 Working load . . . . 24-30 lbs. at 0.70 in.

Clutch separator plates (19)–
Number used .5
Thickness, new . . . . . .0.055-0.0065 in.
Clutch discs (20)–
Number used .4
Thickness, new . . . . .0.071-0.075 in.
Minimum facing groove depth0.006 in.
Spring washers (22)–
Number used, S.N. 015372-025360 . .6
Number used, after 025360 8
Minimum free height . . . . . . .0.126 in.

Install the six brake piston return springs (18) on pins. On early models, position control valve detent (38–Fig. 200) and spring (37) in bore (B). On all models, install backing plate (15–Fig. 193A). Tighten cap screws (15S) to 20 ft.-lbs. torque. When the planet carrier assembly is installed onto brake housing, make sure the planet pinion shafts remain in place.

If screws (3) do not have rubber sleeve molded onto shank, install later type that does. Torque these three screws to 21-25 ft.-lbs.

Use compressed air to check action of both apply pistons. Direct approximately 50 psi to the pressure passage while moving control valve in and out. Standard oil port is diametrically opposed to the control valve. On late models, also test assembly at oil hole located just behind upper part of control valve. Make sure the rear thrust washer (1–Fig. 193A) is in place on planet gear assembly and reinstall the unit, as outlined in paragraph 193.

POWER SHIFT TRANSMISSION

The power shift transmission is available on all models. The power shift transmission provides 8 forward speeds and 4 reverse speeds. Gear changes are accomplished by moving a shift lever and gears can be changed without stopping tractor or operating the foot controlled feathering valve (inching pedal).

OPERATION

Power Shift Models

195. POWER TRAIN. The power shift transmission is a manually controlled, hydraulically actuated planetary transmission consisting essentially of a clutch pack and planetary pack as shown schematically in Fig. 203.

Hydraulic control units consist of three clutch packs (C1, C2 & C3) and four disc brakes (B1 through B4). In addition, a multiple disc clutch (pto) is housed in the clutch pack and used in the pto drive train. All units are hydraulically engaged, and mechanically disengaged when hydraulic pressure to that unit is interrupted. The power train also contains a non-release single disc transmission clutch mounted on engine flywheel, a foot operated inching pedal, a mechanical disconnect for towing and a park pawl.

Three hydraulic control units are engaged for each of the forward and reverse speeds. In 1st speed, Clutch 1 is engaged and power is transmitted to the front planetary unit by the smaller input sun gear (C1S); Brake 1 is engaged, locking the front ring gear to housing, and the planet carrier walks around the ring gear at its slowest speed. Clutch 3 is also engaged, locking the rear planetary unit, and output shaft turns with the planet carrier. Second speed differs from 1st speed only by disengaging Brake 1 and engaging Brake 2, causing planet carrier and output shaft to rotate at a slightly faster speed.

Third speed and 4th speed are identical to 1st and 2nd except that Clutch 1 is disengaged and Clutch 2 engaged, and power enters the front planetary unit through the larger input sun gear (C2S).

Fifth speed and 6th speed differ from 3rd and 4th speeds in the rear planetary unit. Clutch 3 is disengaged and Brake 4 engaged, and the output shaft turns faster than the planet carrier through the action of the rear planet pinions and output sun gear.

In 7th and 8th speeds, both Clutch 1 and Clutch 2 are engaged, locking the input planetary unit, and planet carrier turns with input shaft at engine speed. Engaging the three clutch units locks both planetary units, therefore 7th speed is a direct drive, with transmis

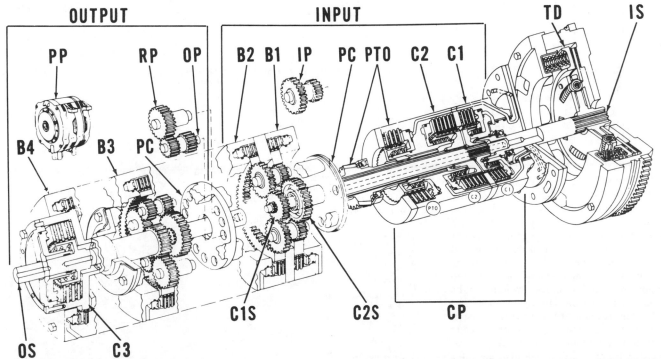

Fig. 203—Schematic view of Power Shift Transmission showing primary function of units. Torsional damper (TD) is a non-release type clutch and pressure plate assembly. The power take-off clutch and drive gear are located in, but are not part of, the transmission power train.

B1. Brake 1	B4. Brake 4	C3. Clutch 3	CP. Clutch pack	OP. Output planet pinion	PP. Planet pack
B2. Brake 2	C1. Clutch	C1S. C1 sun gear	TD. Torsional damper	OS. Output sun gear	PTO. Power take-off drive
B3. Brake 3	C2. Clutch 2	C2S. C2 sun gear	IP. Input planet pinion	PC. Planet carrier	RP. Reverse pinions

sion output shaft turning with, and at the same speed as, the engine. Eighth speed is an overdrive, with transmission output shaft turning faster than engine speed.

Reverse speeds are obtained by engaging Brake 3, which locks the output planetary ring gear to housing, and the output shaft turns in reverse rotation through the action of the two sets of output planetary pinions (RP & OP).

It will be noted that the front planetary unit is an input unit controlled by the two front clutch units in clutch pack and two front brake units in planetary pack. Two input control units must be engaged to transmit power, and five input speeds are obtained by selectively engaging the input brake and clutch units.

The rear planetary unit is an output unit controlled by the two rear brakes and rear clutch. One of the rear control units must be engaged to complete the power train. Two forward ranges and one reverse output range are provided, depending on which rear control unit is engaged.

The accompanying table lists the control units actuated to complete the power flow in each shift position:

	Front (Input) Control Units		Rear (Output) Control Unit
	Forward Speeds		
1st	C1	B1	C3
2nd	C1	B2	C3
3rd	C2	B1	C3
4th	C2	B2	C3
5th	C2	B1	B4
6th	C2	B2	B4
7th	C1	C2	C3
8th	C1	C2	B4
	Reverse Speeds		
1st	C1	B1	B3
2nd	C1	B2	B3
3rd	C2	B1	B3
4th	C2	B2	B3

196. CONTROL SYSTEM. The control valve unit consists of manually actuated speed selector and direction selector valves which operate through four hydraulically controlled shift valves to engage the desired clutch and brake units. The valve arrangement prevents the engagement of any two opposing control units which might cause transmission damage or lockup. Power to operate the transmission system is supplied by an internal gear hydraulic pump mounted on the transmission input shaft, which also supplies the charging fluid for the tractor main hydraulic system. Fluid from the hydraulic pump first passes through a full flow oil filter to the main transmission

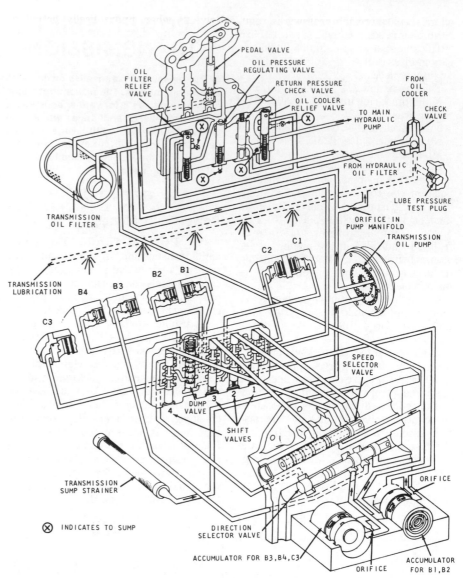

Fig. 204—Schematic view of the oil control circuits, valves and accumulators in the Power Shift transmission.

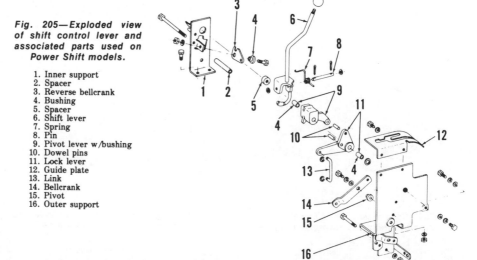

Fig. 205—Exploded view of shift control lever and associated parts used on Power Shift models.

1. Inner support
2. Spacer
3. Reverse bellcrank
4. Bushing
5. Spacer
6. Shift lever
7. Spring
8. Pin
9. Pivot lever w/bushing
10. Dowel pins
11. Lock lever
12. Guide plate
13. Link
14. Bellcrank
15. Pivot
16. Outer support

oil gallery, where the pressure is regulated at approximately 175 to 195 psi for the transmission control functions. Excess oil passes through the regulating valve to the oil cooler and main hydraulic pump.

Fluid from the transmission main oil gallery is routed through the inching pedal valve to Clutches 1 and 2; and through a spring-loaded accumulator to the brake actuating pistons and to Clutch 3.

Refer to Fig. 204 for a schematic view of control circuits. The direction selector valve and shift valve 4 controls the routing of pressure to the output control units (Clutch 3, Brake 3 & Brake 4). The speed selector valve contains four pressure ports (1, 2, 3 & 4) which control the movement of the four shift valves by pressurizing the closed end (opposite the return spring) when port is open to pressure. Neutral position is provided by the selector valves or by depressing the inching pedal.

When the direction selector valve is moved to the forward detent position, system pressure is routed in Shift Valve 4. In the low range positions (1st through 4th gears), the speed selector valve charging port to the top of Shift Valve 4 is open to pressure, Shift Valve 4 moves downward against spring pressure and Clutch 3 is actuated. In the high range positions (except 7th gear which is direct and uses all three clutch units), charging pressure is cut off to Shift Valve 4, shift valve return spring moves valve upward and Brake 4 is actuated. When the direction selector valve is moved to reverse detent position, system pressure by-passes Shift Valve 4 and is routed directly to Brake 3. In the neutral detent position, system pressure is cut off from all three output control units.

Shift Valves 1, 2 & 3 direct system pressure to the input control units (Clutches 1 & 2 and Brakes 1 & 2). Shift Valve 1 directs pressure to Clutch 2 when hydraulically actuated and to Clutch 1 when charging port to top of Valve 1 is closed.

Shift Valve 2 routes pressure to Shift

Valve 3 when hydraulically actuated, and permits simultaneous engagement of Clutches 1 and 2 when charging port to top of Valve 2 is closed.

Shift Valve 3 directs pressure to Brake 2 when hydraulically actuated and to Brake 1 when charging port to top of Valve 3 is closed.

197. LINKAGE ADJUSTMENT. To adjust the shift control linkage, refer to Fig. 206. Disconnect direction selector control cable yoke from arm. Move shift lever to "NEUTRAL" position. With direction selector arm in the center of three detent positions (Neutral), adjust length of control cable until pin hole is aligned in bellcrank and yoke. Be sure shift lever is centered, as if to shift into "PARK". Install pin, shift to "PARK" position and check for interference. Make minor adjustments as required.

With speed selector cable yoke disconnected from arm on lever, move shift lever to "NEUTRAL" position and place speed selector arm in uppermost detent position. Adjust yoke if neces-

Fig. 207—Pedal valve adjustment point on Power Shift models.

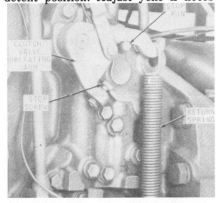

sary, until connecting pin can just be inserted. Check PARK lock cable and lever to be sure the cable clamp nut on right side of transmission housing is tight, and that park pawl fully engages when shifted to PARK. Adjust by moving the lock cable swivel at the upper end of cable.

To adjust pedal valve linkage, refer to Fig. 207. Remove clutch pedal return spring. Loosen locknut on stop screw and run screw in. Turn valve operating arm fully counter-clockwise until valve bottoms and adjust stop screw until screw head touches stop pin. Back out one-half turn and tighten locknut.

To adjust pedal height, measure from footrest to the lowest part of pedal. Distance should be 5-1/8 inches (Fig. 208). Adjust by removing hex adapter from pedal, rotate until the desired position is obtained and reinstall adapter.

198. PRESSURE TEST AND ADJUSTMENT. Before checking the transmission operating pressure, first be sure that transmission oil filter is in good condition and that oil level is at top of "SAFE" mark on dipstick. Place towing disconnect lever in "TOW" position, start engine and operate at 1900 rpm until transmission oil is at operating temperature. If tractor is equipped with Power Front Wheel Drive, put main hydraulic pump out of stroke by turning in the shut-off screw.

Stop engine and install a 0-300 psi pressure gage in "SYSTEM PRESSURE" plug hole (Fig. 209). Gage should register 175-195 psi with engine operating at 1900 rpm and speed control lever in any position. If pressure is not as indicated, remove plug and add or remove shims (12—Fig. 218 or Fig. 219) located between plug and spring, until pressure is correct.

If pressure cannot be adjusted, other possible causes are:

1. Incorrect pedal valve linkage adjustment, refer to paragraph 197.

2. Malfunctioning regulator valve, pedal valve or oil filter relief valve;

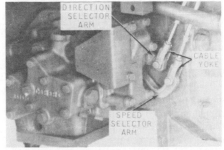
Fig. 206—Power shift linkage adjustment yokes and selector arms.

Fig. 208—Remove hex adapter and rotate to obtain pedal height as shown.

Fig. 209—View of control valve housing showing pressure gage points.

overhaul as outlined in paragraphs 209 or 210.

If adjustment of operating pressure does not correct the malfunction, check pressures as outlined in paragraph 199.

TROUBLE SHOOTING

Power Shift Models

199. **PRESSURE TEST.** To make a complete check of transmission hydraulic system pressures, first check and adjust operating pressure as outlined in paragraph 198, then proceed as follows:

Install four 0-300 psi pressure gages in the four plug holes indicated in Fig. 209 at C-1, C-2, (B-1, B-2) and C-3 (B-3, B-4). Have engine running at 1900 rpm, operating temperature up to normal and towing disconnect lever in "TOW" position. As the shift lever is placed in each position of the following table, observe pressures in the appropriate gages for each element as listed.

Neutral		B1	
Neutral Fwd.		B1	C3
1st Fwd.	C1	B1	C3
2nd Fwd.	C1	B2	C3
3rd Fwd.	C2	B1	C3
4th Fwd.	C2	B2	C3
5th Fwd.	C2	B1	B4
6th Fwd.	C2	B2	B4
7th Fwd.	C1 C2	* *	C3
8th Fwd.	C1 C2	* *	B4
Neutral Rev.		B1	B3
1st Rev.	C1	B1	B3
2nd Rev.	C1	B2	B3
3rd Rev.	C2	B1	B3
4th Rev.	C2	B2	B3

7th and 8th speed check shows pressure on gage at B1, B2 pressure port but elements are not engaged since pressure is stopped at accumulator.

Excessive leakage is indicated if pressures are more than 15 psi below system pressure in B-1, B-2, B-3, B-4 or C-3, or more than 25 psi below in C-1 or C-2.

NOTE: Because of line restriction, C-1 and C-2 pressure is approximately 10 psi below system pressure in normal operation.

With shift lever in any forward or reverse gear and engine speed at 1900 rpm, depress the inching pedal while noting C-1 or C-2 pressure gage reading. Gage pressure on either or both gages showing pressure should drop to zero with pedal fully depressed. Release the pedal slowly; gage pressure should rise at a smooth, even rate until approximately 80 psi is registered with pedal ½ to 1 inch from top; then move quickly to operating pressure

with further pedal movement.

Failure to perform as outlined would indicate maladjustment of pedal valve linkage (see paragraph 197) or malfunction of the valve (overhaul as outlined in paragraph 209).

200. **LUBRICATION PRESSURE TEST.** Before overhauling or repairing the Power Shift transmission, a lubrication pressure test can help isolate the problem, and prove the results after repair.

The lubrication pressure plug to be removed for the test is located in clutch housing on upper right side, just to the right of oil return line. It will be necessary to remove the footrest panel to reach the plug for this test. Install a 0-60 psi gage that does not contain a dampener orifice. In PARK position and with engine at operating temperature, the pressure should be at least 30 psi at 2200 rpm. Excessive pressure indicates a blockage in the lube circuit. If pressure is low, put the main hydraulic pump out of stroke by turning in the shut-off screw. If pressure comes up to normal, there is a leak in the hydraulic system. Next run engine at a slow enough speed to get a reading of 5-10 psi. Place the towing lever in TOW position. Shift the transmission through the positions shown in table in paragraph 199 while observing pressure gage. Pressure should drop momentarily, then come back to within 3 psi of original value. If pressure fails to return at any selected speed, depress clutch pedal to cut off pressure to C-1 and C-2. If pressure comes back up, a clutch unit is leaking. If it does not come up with clutch depressed, the planetary pack is leaking. Test in several speeds to pinpoint leaking unit while consulting shift table.

Engage pto clutch and observe gage. If pressure fails to return to within 3 psi of original pressure, the pto clutch is leaking or the valve is out of adjustment. Refer to paragraph 206 for overhaul of pto clutch or paragraph 258 for pto valve adjustment.

201. **BEHAVIOR PATTERNS.** Erratic behavior patterns can be used to pinpoint some systems malfunctions.

ODD SHIFT PATTERN. If tractor slows down when shifted to a faster speed; speeds up when shifted to a slower speed; or fails to shift when selector lever is moved; a sticking shift valve is indicated. Refer to paragraphs 196 and 199. Overhaul the control valve as outlined in paragraph 211.

SLOW SHIFT. Possible causes are; improper regulating valve adjustment; improper pedal linkage adjustment;

plugged fluid filter; malfunctioning regulating valve, pedal valve or oil filter relief valve; broken accumulator spring; sticking accumulator piston; or slipping clutch or brake unit or units.

ROUGH PEDAL ENGAGEMENT. If tractor jumps rather than starts smoothly when pedal valve is actuated, a sticking clutch pedal valve or broken pedal valve spring is indicated.

SLIPPAGE UNDER LOAD. If transmission slips, partially stalls or stalls under full load, first check the adjustment of pedal valve as outlined in paragraph 197, then check transmission pressures as outlined in paragraphs 198 and 199. If trouble is not corrected, one of the clutch units; or torsional damper is malfunctioning. Refer to paragraph 212 for torsional damper overhaul.

If a clutch or brake unit is suspected of slipping, it will be necessary to determine which of the three units is at fault in that speed. Refer to the table in paragraph 199 to determine which three units are involved in the speed range in question. Then prove one unit at a time by choosing a speed that utilizes that particular unit, and if that speed does not slip, choose another speed that changes only one unit whenever possible. In this way the slipping unit can be isolated.

NOTE: 4th to 5th and 5th to 4th change two units at once, as do 6th to 7th and 7th to 6th.

If clutch C1 or C2 is suspected, the inching pedal can be used to determine if either clutch is bad. Since the only difference between 2nd and 4th speed is the change from C1 to C2, these two speeds can be used to isolate the faulty unit. (If either clutch is slipping, it will also slip in 7th and 8th speeds since both clutches are applied in both speeds). Place shift lever in speed desired and allow brake and clutch units time to engage before releasing inching pedal.

If one or more units are found to be slipping in every gear in which the unit is engaged, remove and overhaul the transmission as outlined in paragraphs 204 through 216.

TRACTOR FAILS TO MOVE. If tractor fails to move when transmission is engaged, first check to see that tow disconnect is fully engaged. If tow disconnect unit is engaged, check to see that park pawl operates properly and is correctly adjusted. Park pawl is engaged by cam action and disengaged by a return spring. If spring breaks or becomes unhooked, pawl may remain engaged even though linkage operates satisfactorily. To examine or renew the park pawl return spring, remove transmission housing cover as outlined in paragraph 217.

TRACTOR CREEPS IN NEUTRAL. A slight amount of drag is normal in the clutch and brake units, especially when transmission oil is cold. Excessive creep is usually caused by warped clutch or brake plates; observe the following.

If tractor creeps when inching pedal is depressed and properly adjusted, either Clutch 1 or Clutch 2 is malfunctioning. Check as follows: With engine speed at 1500 rpm and transmission fluid at operating temperature shift to 2nd speed on a flat surface. Depress the inching pedal; if tractor continues to roll forward at approximately the same speed, Clutch 1 is malfunctioning, if tractor speed increases, Clutch 2 is dragging.

Place shift lever in NEUTRAL position. Disconnect yoke from direction selector arm on transmission (Fig. 206). This will leave the output section of transmission in neutral. Shift into 1st forward with shift lever, but do not depress clutch pedal. If tractor creeps forward with throttle set at 1500 rpm

and transmission oil at operating temperature, Clutch 3 or Brake 4 is dragging; if tractor creeps backward, Brake 3 is malfunctioning.

NOTE: Dragging clutch or brake units, aside from causing creep, will contribute to loss of power, heat and excessive wear. Creep is merely an indication of possibly more serious trouble which needs to be corrected for best performance, or to prevent future failure.

REMOVE AND REINSTALL

Power Shift Models

202. **PEDAL AND REGULATING VALVES.** Pedal valve housing and regulating valve housing attach to left side of clutch housing using common gaskets and gasket plate. Housings may be removed separately, but both should be removed to renew the gaskets.

To remove the housings, remove left battery and battery box if necessary. Remove pedal return spring and disconnect rod from valve operating arm. Disconnect lube pipe, inlet pipe and

outlet pipe from housings. Disconnect pto valve operating rod at clutch housing, then unbolt and remove the housings, gasket plate and gaskets.

Overhaul the pedal valve housing as outlined in paragraph 209 and regulating valve housing as in paragraph 210.

When installing, use light, clean grease to position gaskets and gasket plate, making sure gaskets are installed on proper sides of plate as shown in Fig. 218. Install regulator valve housing and retaining cap screws, then install pedal valve housing. Tighten retaining cap screws evenly and securely, and complete the assembly by reversing the disassembly procedure. Adjust as outlined in paragraphs 197 and 198.

203. **CONTROL AND SHIFT VALVES.** To remove the control and shift valve housing, drain transmission and remove right battery and battery box if necessary. Disconnect wiring to start-safety switch and remove the cotter pins which retain control cable yokes to control arms. Remove the retaining cap screws, then remove valve housing, being careful not to lose the detents and springs (14, 15 & 16—Fig. 220).

Overhaul the removed unit as outlined in paragraph 211 and install by reversing the removal procedure. Make sure the inner and outer gaskets are in proper order since they are not interchangeable. Tighten retaining cap screws evenly and securely. Adjust as outlined in paragraph 197.

204. **TRACTOR SPLIT.** To obtain access to engine flywheel, torsional damper, pto drive gears, or power shift transmission main components, it is first necessary to detach (split) clutch housing from engine block; proceed as follows:

Drain cooling system and remove hood and muffler. Remove side shields, grille screens and right and left cowl side covers. Remove step plates and battery boxes.

Discharge brake accumulator by opening the right brake bleeder screw and holding brake pedal down in a few moments.

If equipped with air conditioning, use two wrenches to break the two connections on the lines on left side of tractor, just below cab. If gas can be heard leaking, tighten and reloosen connection until coupler will part without leaking. Cap connections to keep out dirt.

Disconnect throttle rod, wiring, hydraulic lines, tachometer cable and temperature indicator sending unit. Be

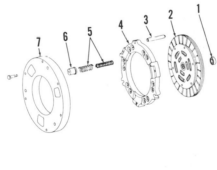

Fig. 210—Install jack screws in the square openings in early model pressure plate.

1. Pilot bearing	3. Pin (3 used)	5. Springs
2. Torsional damper disc	4. Pressure plate	6. Spring cup
		7. Cover

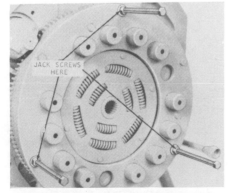

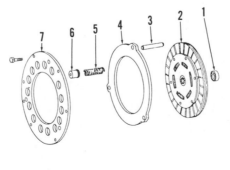

Fig. 211—In late model pressure plate, install jack screws ONLY in holes shown.

1. Pilot bearing	3. Pin (3 used)	5. Spring
2. Torsional damper disc	4. Pressure plate	6. Spring cup
		7. Cover

sure to cap all disconnected hydraulic fittings to prevent dirt entry. If tractor is equipped with front end weights, remove the weights. If equipped with Power Front Wheel Drive, disconnect inlet and outlet oil pipes. Support engine and transmission securely and separately, remove the connecting cap screws and roll transmission assembly rearward away from engine.

To attach, reverse the above procedure and tighten the connecting cap screws to a torque of 170 ft.-lbs. on 5/8-inch screws, or 300 ft.-lbs. on ¾-inch screws.

205. **TORSIONAL DAMPER.** Power Shift tractors are equipped with a non-release type clutch pressure plate, and a disc assembly with springs mounted radially around the disc hub, which absorbs torsional loads in the same manner as does a conventional dry clutch. This action softens the shock of initial engagement at a standstill, and when speed changes are made while in motion.

To remove torsional damper, detach clutch housing from engine as outlined in paragraph 204. Early tractors were equipped with a flat pressure plate (Fig. 210), which used inner and outer springs (except 4230 models) for pressure application. Late tractors use a different type pressure plate which has the spring retainers protruding through the cover (Fig. 211) and is equipped with single springs.

On early type pressure plates, install three ½-inch jack screws, nuts and large washers in the three square openings as shown in Fig. 210. After jack screws are threaded into cover and locknut tightened, remove the six cover cap screws. Cover can now be

removed as a unit, and reinstalled without disassembly if desired, or back off the locknuts on jack screws to disassemble. The torsional disc may have a single or a double row of springs around the hub. Reinstall in reverse order of removal, making sure the short side of disc hub faces out. Use a suitable disc aligning tool to center disc before tightening cover cap screws to 35 ft.-lbs. torque. Remove jack screws.

On the late type pressure plate (Fig. 211), remove the three cover cap screws which are located at the center of the three spring retainer groups. Install three jack screws ONLY IN HOLES SHOWN to prevent warping the cover by uneven pressure, which could happen if jack screws were installed in holes *between* spring retainer groups. Remove remaining cap screws after locknuts are tightened on jack screws. Cover, springs and retainers can be removed from flywheel by backing off the three locknuts to relieve spring pressure. (No further disassembly is necessary). Reinstall in reverse order of removal, with a suitable aligning tool in disc and short side of hub out. Install jack screws, run the nuts down until three pressure plate cap screws can be installed, then remove jack screws and install remaining cap screws. Tighten to 35 ft.-lbs. torque.

206. **CLUTCH PACK AND TRANSMISSION PUMP.** The transmission pump and clutch pack can be removed as a unit after draining hydraulic system and detaching clutch housing from engine as outlined in paragraph 204.

Overhaul the removed clutch pack and pump as outlined in paragraphs 213 and 214.

To assist in easier installation of clutch pack, use alignment studs in the two side holes of housing and position gasket using light grease. Make sure oil passages in clutch housing and gasket are properly aligned. Insert connecting shafts with sealing ring (5) to rear. Install clutch pack with oil passages in mounting flange aligned with those of gasket and clutch housing. Tighten cap screws to 35 ft.-lbs. torque.

207. **CLUTCH HOUSING.** The clutch housing must be removed for access to pto drive gear train or removal of planetary unit. To remove the clutch housing, first split tractor between engine and clutch housing as outlined in paragraph 204 and remove clutch pack as in paragraph 206.

Remove Sound-Gard body, or 4 post Roll-Gard if so equipped. Remove batteries and battery boxes, differential lock pedal and front platform.

Mark and diagram location of hydraulic tubing if necessary, then remove hydraulic tubes and system control linkage. Remove rockshaft housing from tractor as outlined in paragraph 282.

Remove the snap ring (Fig. 213) securing pto clutch gear bearing to bore

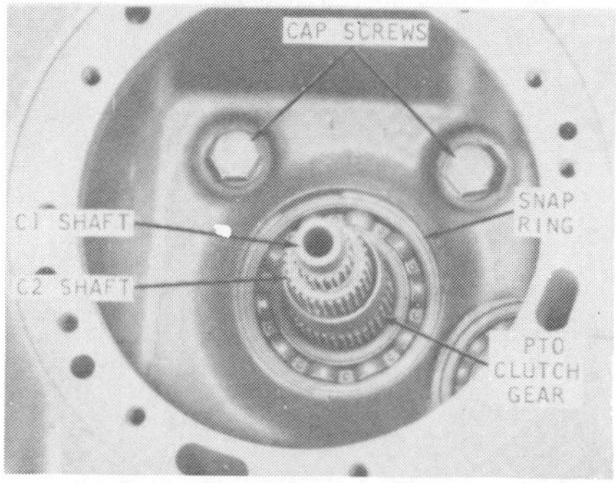

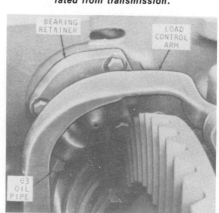

Fig. 213—View of clutch housing with clutch pack removed. Snap ring, pto clutch gear and the two hidden cap screws must be removed before clutch housing can be separated from transmission.

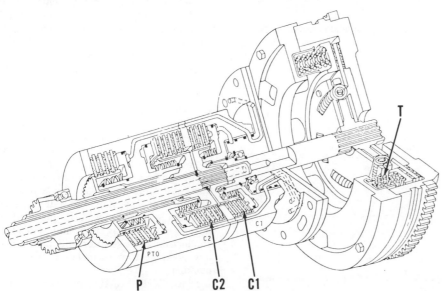

Fig. 212—Cross-section showing early type torsion damper (T), front clutch (C1), second clutch (C2), pto clutch (P) and associated parts. Other models are similar.

Fig. 214—Loosen cap screws to load control arm support for more wrench clearance to C3 oil pipe fitting.

and withdraw clutch gear and bearing as a unit.

NOTE: A slide hammer may be required to remove the gear.

Support clutch housing and steering support assembly from a hoist and remove clutch housing flange cap screws. Two upper, center screws are accessible through inside front of clutch housing as shown in Fig. 213. Pry clutch housing from its doweled position on transmission case and swing housing away from rear unit.

Use new gasket and "O" rings when reinstalling clutch housing. "O" rings may be held in position with grease. Tighten 5/8-inch cap screws to a torque of 170 ft.-lbs. and ¾-inch cap screws to 300 ft.-lbs.

208. PLANETARY PACK. To remove the transmission planetary pack, first detach (split) engine from clutch housing as outlined in paragraph 204; remove clutch pack as in paragraph 206 and clutch housing as in paragraph 207. Remove rockshaft housing as in para-

graph 282 and transmission top cover plate.

Disconnect Clutch 3 pressure tube from output reduction gear rear bearing retainer and remove retainer (Fig. 214). If necessary to obtain more clearance to C3 tube fitting, loosen the cap screws holding the load control arm support at the bottom of housing. Using a long brass drift and reaching through center of planetary pack from front, drive reduction gear rearward until bearing cup is removed from rear bore. Withdraw shaft as shown in Fig. 215.

Remove the four retaining cap screws and, using a hoist and suitable lifting fixture, lift out planetary pack as shown in Fig. 216. Overhaul the removed planetary unit as outlined in paragraph 215 and 216.

Before installing planetary pack, inspect or renew the four brake-passage "O" rings in bottom of transmission housing. Reduction gear shaft should

be installed with 0.000-0.002 inch bearing preload. To check bearing adjustment, make a trial installation of shaft, bearing cup and retainer, using one additional shim between bearing retainer and housing. Tighten retaining screws securely and measure shaft end play using a dial indicator. Remove the retainer and deduct from shim pack, shims equal to the measured end play plus 0.001 inch. Keep remainder of shim pack together for final installation.

Lower planetary unit straight downward being careful not to dislodge brake passage "O" rings. Tighten the four retaining cap screws alternately and evenly to 55 ft.-lbs. torque, and complete the assembly by reversing the disassembly procedure.

OVERHAUL
Power Shift Models

209. PEDAL VALVE. Refer to Fig.

Fig. 215—Remove bearing retainer and C3 oil pipe and withdraw reduction gear shaft before attempting to lift off planetary pack. 4630 model shown; other models have ring gear on opposite side.

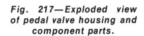

Fig. 217—Exploded view of pedal valve housing and component parts.

1. Housing
2. Shaft (Clutch)
3. Special pin (2)
4. Link (2)
5. Shaft (pto)
6. Arm (pto valve)
7. Clutch valve shaft
8. Shim
9. Spring (4)
10. Clutch valve
11. Spring pin (2)
12. Special plug
13. Pto valve
14. Pto valve shaft

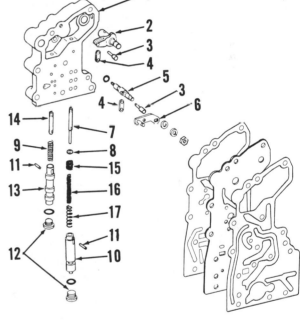

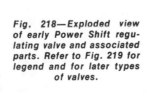

Fig. 216—Use a suitable lifting fixture and hoist to remove planetary pack.

Fig. 218—Exploded view of early Power Shift regulating valve and associated parts. Refer to Fig. 219 for legend and for later types of valves.

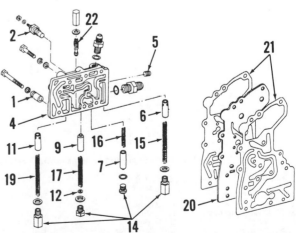

217 for an exploded view of pedal valve and associated parts. Refer to paragraph 202 for removal and installation information.

Valve spools (10 & 13) must slide smoothly in their bores and must not be scored or excessively loose. Refer to Fig. 217 and check the pto and pedal valve springs for distortion and against the values which follow:

Pto Valve Spring (9)
   Free Length . . . . .1.8 in.
   Lbs. Test at Inches . . .16.6-20.4 at 1.4
Pedal Valve Lower Spring (15)
   Free Length . . . . .0.5 in.
   Lbs. Test at Inches . . . .3.3-4.1 at 0.34
Pedal Valve Center Spring (17)
   Free Length . . . . .1.3 in.
   Lbs. Test at Inches . . .15.1-18.3 at 1.1
Pedal Valve Upper Spring (16)

Free Length . . . . .2 in.
Lbs. Test at Inches . . .11.7-14.3 at 1.4

210. REGULATING VALVE. Refer to Fig. 218 or 219 for an exploded view of regulating valve and associated parts and to paragraph 202 for removal and installation information.

Spools for oil cooler relief valve (6—Fig. 218) and filter relief valve (11) are interchangeable, and return pressure check valve (7) and system pressure regulating valve (9) are interchangeable, but valve springs must be marked or tested for proper installation. All valves except check valve (7) in Fig. 219 are interchangeable, but the springs are not. Late model regulating valves (Fig. 219) no longer use the hydraulic oil by-pass screw (22—Fig. 218), but are equipped with "O" rings on all valves except the return pressure check valve (7—Fig. 219). A transmission lube valve (10) has been added to the late model unit. Check the valves and bores for sticking or scoring. Valves must move freely in bore without excessive clearance. Refer to appropriate Fig. 218 or 219 and check the springs against the values which follow:

Early Type Fig. 218
Oil Cooler Relief Valve Spring (15)
   Color . . . . .Green
   Free Length . . . . .5.75 in.
   Lbs. Test at Inches . . .33.1-40.5 at 4.5

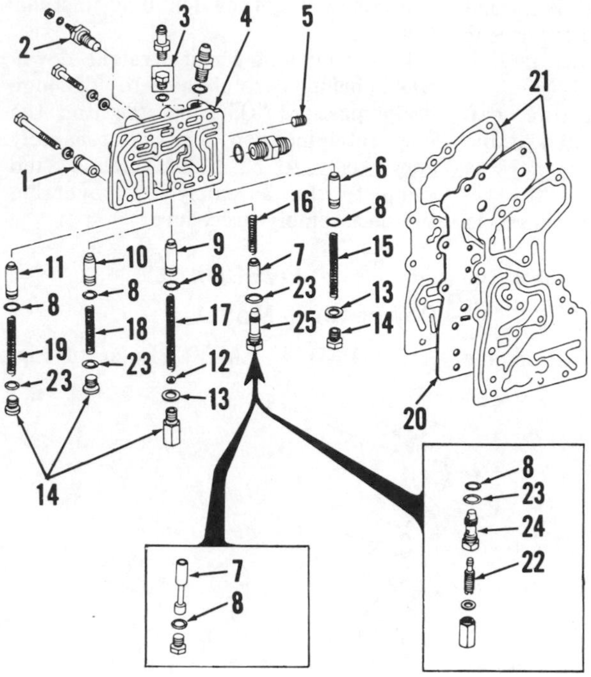

Fig. 219—Exploded view of late Power Shift regulating valve and associated parts. Parts in insets are used in earlier valves. Refer to Fig. 218 for earliest type valve.

1. Spacer
2. Temperature sender
3. Plug
4. Housing
5. Plug
6. Cooler relief valve
7. Return check valve
8. "O" ring
9. Regulating valve
10. Lube oil valve
11. Filter relief valve
12. Adjusting shim
13. Washer
14. Plugs
15. Spring
16. Spring
17. Spring
18. Spring
19. Spring
20. Plate
21. Gaskets (different)
22. By-pass screw
23. "O" ring
24. Plug
25. Plug

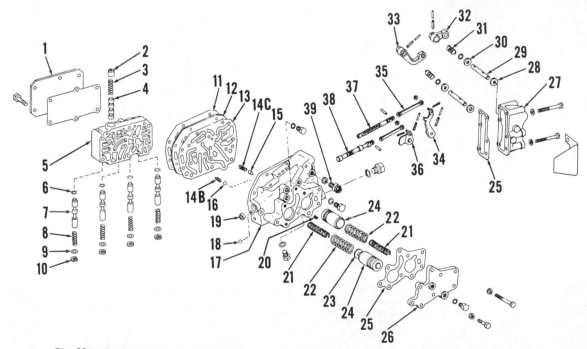

Fig. 220—Exploded view of control valve assembly and associated parts used on Power Shift models.

1. Cover
2. Plug
3. Spring
4. Dump valve
5. Shift valve housing
6. "O" ring
7. Shift valve
8. Spring
9. Washer
10. Retaining ring
11. Gasket
12. Plate
13. Gasket
14B. Spring for ball
14C. Spring for plunger cup
15. Detent cone (2)
16. Ball (2)
17. Control valve housing
18. Ball
19. Plug
20. Orifice (2)
21. Spring (inner)
22. Spring (outer)
23. Special washer
24. Accumulator piston
25. Gasket
26. Piston cover
27. Linkage cover
28. Oil seal (2)
29. Shaft
30. Felt washer (2)
31. Spring (2)
32. Reverse arm
33. Speed control arm
34. Operating arm
35. Link
36. Operating arm
37. Speed control valve
38. Direction valve
39. Start safety switch

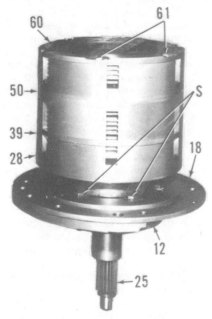

Return Pressure Check Valve
Spring (16)
    Free Length . . . . .2.31 in.
    Lbs. Test at Inches . .10.8-13.2 at 0.75
Pressure Regulating Valve Spring (17)
    Free Length . . . . .3.94 in.
    Lbs. Test at Inches . .58.5-71.5 at 3.44
Oil Filter Relief Valve Spring (19)
    Color . . . . .Yellow
    Free Length . . . . .5.9 in.
    Lbs. Test at Inches . .23.2-28.4 at 4.31

Later Types Fig. 219
Oil Cooler Relief Valve Spring (15)
    Free Length . . . . .4.55 in.
    Lbs. Test at Inches . .35.7-43.7 at 3.53
Return Pressure Check Valve
Spring (16)
    Free Length . . . . .2.31 in.
    Lbs. Test at Inches . .10.8-13.2 at 0.75
Pressure Regulating Valve Spring (17)
    Free Length . . . . .3.94 in.
    Lbs. Test at Inches . .58.5-71.5 at 3.44
Transmission Lubrication Oil Valve
Spring (18)

Free Length . . . . .2.83 in.
    Lbs. Test at Inches . .11.7-14.3 at 2.13
Oil Filter Relief Valve Spring (19)
    Free Length . . . . .3.74 in.
    Lbs. Test at Inches . .24.3-29.7 at 2.61

211. **CONTROL VALVE.** The shift valve housing is attached to inner face of control valve housing by a cover and six cap screws. Refer to Fig. 220 for an exploded view. Refer to paragraph 203 for removal and installation procedure.

To disassemble the removed unit, remove the six cap screws retaining shift valve housing (5) and lift off cover (1), housing, gasket plate (12) and gaskets (11 & 13). Note that the two gaskets are not interchangeable. Mark the removed gaskets "Outer" and "Inner" as they are removed to aid in the installation of new gaskets when unit is reassembled. Lift out inner detent springs (14B & 14C), plungers (15) and balls (16) to prevent loss as

Fig. 221—View of removed clutch assembly. The input shaft (25) and release bearing sleeve can be inserted into hole in bench or holding fixture to facilitate disassembly and reassembly. Screws (S) attach manifold plate (18) and pump housing (12) together. Refer to Fig. 222 for legend.

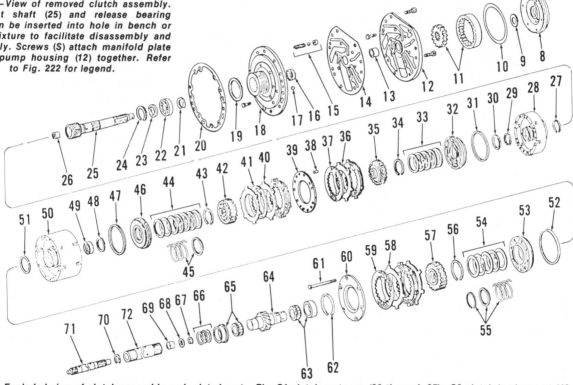

Fig. 222—Exploded view of clutch assembly and related parts. The C1 clutch parts are (28 through 37); C2 clutch is shown at (40 through 48); the pto clutch is at (51 through 59).Coil spring, cup and snap ring shown at (45 and 55) are used in place of Belleville washer springs shown at (44 and 54) on some models. Parts shown are typical but differences may be noted.

8. Pump body	25. Input shaft	37. Clutch friction discs (alternate with plates 36)
9. Seal	26. Bushing	
10. Packing ring	27. Snap ring (same as 29)	38. Dowel pin
11. Pump gears		39. Clutch separator plate
12. Pump housing	28. Front clutch drum	
13. Bushing	29. Snap ring (same as 27)	40. Clutch plates (flat plate next to piston 46)
14. Gasket		
15. Spring, ball and plug	30. Piston inner seal	41. Clutch friction discs (alternate with plates 40)
16. Bushing	31. Piston outer seal	
17. Steel balls (2 used)	32. Low range clutch piston	42. Clutch hub
18. Clutch manifold	33. Belleville spring washers	43. Snap ring
19. Sealing ring (4 used)		44. Belleville spring washers
20. Gasket	34. Snap ring	
21. Snap ring	35. Snap ring	45. Coil spring and cup (some models in place of 44)
22. Ball bearing	36. Clutch plates (flat plate next to piston 32)	
23. Thrust washer		46. High range clutch piston
24. Snap ring		47. Piston outer seal

	48. Piston inner seal	63. Bearing assembly (ball bearing 4230, 4430, 4630 models)
	49. Bushing	
	50. Rear clutch drum	64. Pto drive shaft and gear
55. Coil spring, cup and snap ring (some models in place of 54)	51. Piston inner seal	
	52. Piston outer seal	65. Bearing assembly (ball bearings 4230, 4430, 4630 models)
56. Snap ring	53. Pto clutch piston	
57. Clutch hub	54. Belleville spring washers	66. Roller thrust bearing and races (some models)
58. Clutch plates (flat plate next to piston 53)		
59. Clutch friction discs (alternate with 58)		67. Snap ring
60. Clutch backing plate		68. Thrust washer
61. Through bolts		69. Bushing
62. Snap ring (variable thickness on models with taper roller bearings at 63 and 65)		70. Snap ring (4630 model)
		71. Low range (C1) clutch shaft
		72. High range (C2) clutch shaft

housing (17) is removed. Invert control valve housing (17) and remove the eight cap screws retaining accumulator cover (26), gasket (25), pistons (24) and springs (21 and 22). Note that the accumulator pistons face opposite each other, and washer (23) is installed behind the outward facing piston only. Remove the six cap screws retaining cover (27) to housing and withdraw cover, control valves and operating linkage as a unit from control valve housing. Remove plug (2) from shift valve housing and withdraw spring (3) and dump valve spool (4). Remove the four retaining rings (10) and washers (9) then withdraw shift valve springs (8) and spools (7). The four shift valve spools and springs are interchangeable. The two replaceable accumulator charging orifices can be removed at this time.

Spring (14B) for the reverse detent ball (16) should have free length of 1.12 inch and should exert 14.4-17.6 pounds force when compressed to height of 0.80 inch. Spring (14C) for the speed detent cone (15) is similar in size, but is different. Spring (14C) should have 1.16 inch free length and should exert 5.85-7.15 pounds force when compressed to 0.83 inch.

Clean all parts in a suitable solvent and check for scoring or other damage, and for free movement of valve spools in bores. Control valve actuating mechanism need not be disassembled unless renewal of parts is indicated.

Check the valve springs for damage or distortion and against the values which follow:
Dump Valve Spring (3—Fig. 220)
    Free Length . . . . 1.12 in.
    Lbs. Test at Inches . . 31.5-38.5 at 0.84
Shift Valve Spring (8)
    Free Length . . . . 1.39 in.
    Lbs. Test at Inches . . . . 6.8-8.4 at 0.81
Accumulator Valve Outer Spring (22)
    Free Length . . . . 4.09 in.
    Lbs. Test at Inches . . . 185-225 at 2.91
                       52-65 at 3.76
Accumulator Valve Inner Spring (21)
    Free Length . . . . 3.78 in.
    Lbs. Test at Inches . . . 100-122 at 2.91

Reassemble by reversing the disassembly procedure, using Fig. 220 as a guide. Tighten the six cap screws retaining shift valve housing to control valve evenly and alternately to a torque of 20 ft.-lbs. and tighten accumulator cover screws to 35 ft.-lbs. torque.

212. TORSIONAL DAMPER. To overhaul torsional damper, remove unit as outlined in paragraph 205 and determine if it is an early or late model by referring to Fig. 210 or 211. Check the

removed disc for loose rivets, cracks or broken springs. Facing should be smooth and free of grease or oil. Renew disc if facing grooves are worn away, and determine the cause of wear, since this type coupler is never released and should not slip to any appreciable degree.

Friction surface of flywheel must not be heat-checked or scored, or have more than 0.006 inch out-of-true irregularities. If machining is necessary, do not remove more than 0.060 inch material, or flywheel must be renewed. If in doubt as to the amount of material that has been removed, check the depth of friction surface from the pressure plate mounting flange and compare to a new unit.

Pins in flywheel should protrude 0.43 inch past clutch cover mounting surface of flywheel. Tighten screws retaining flywheel to crankshaft to 85 ft.-lbs. torque.

The pilot bearing in flywheel is sealed and cannot be lubricated. Renew bearing if loose, rough or dry.

The pressure plate drive pin notches must not be excessively worn, and the friction surface must not be scored or more than 0.006 inch out-of-true. Renew pressure plate if any of these conditions are found. Check springs for rust, pitting or distortion, and against the test values which follow:

Early Type (Fig. 210)
Number of Springs Used—
    4230 . . . . 12 Outer
    4430 . . . . 12 Inner & 12 Outer
    4630 . . . . 15 Inner & 15 Outer
Spring Free Length—
    4230 . . . . 2.91 in.
    4430 & 4630 Outer Spring . . . . 2.91 in.
4430 & 4630 Inner Spring . . . . . . 3.22 in.
Lbs. Test at Inches
    4230 . . . . 153-187 at 1.81
    4430 & 4630 Outer
      Spring . . . . 153-187 at 1.81
    4430 & 4630 Inner
      Spring . . . . 44-55 at 1.81

Late Type (Fig. 211)
Number of Springs Used—
    Late 4230 9
    Late 4430, late 4630 12
Lbs. Test at Inches—
    Late 4230, late 4430 . . . 198-242 at 1.6
    Late 4630 . . . . 247-303 at 1.66

To reassemble unit, refer to paragraph 205 and note proper holes to use when reinstalling jack screws on late type pressure plate. Tighten pressure plate retaining screws to 35 ft.-lbs. torque.

213. CLUTCHES. To disassemble the

removed clutch pack, use a holding fixture with a 2-inch hole (or drill a 2-inch hole near edge of a table or bench). Insert input shaft (25—Fig. 221) and release bearing sleeve through hole, with pto clutch pressure plate (60) up. Remove through-bolts (61) then lift off pto clutch pressure plate, pto clutch discs and clutch hub. Pto and C2 clutch drum (50), C1 and C2 pressure plate (39) and C1 clutch drum (28) can be separated after jarring slightly.

C1 clutch drum can be lifted off input shaft and manifold assembly after removing the snap ring (29) on rear splines of input shaft (25).

Clutch plates (36, 40 & 58—Fig. 222), friction discs (37, 41 & 59) and other parts may be interchangeable between certain locations on some models; however, be sure to separate and identify by location all parts as they are separated to prevent improper assembly. Refer to Fig. 222 and check parts against the values which follow:

4230
C1 Clutch
    Piston (32)—I.D. . . . . . . 3.125-3.126 in.
      O.D. . . . . 6.600-6.610 in.
    Springs (33)—No. used 4
    Minimum free height of
      one spring . . . . 0.120 in.
    Drive plates (36)—
      No. used—Wavy . . . 2 (Notched tab)
      Flat . . . . 1 (Not notched)
      Thickness—Wavy . . 0.118-0.122 in.
      Flat . . . . 0.115-0.125 in.
    Friction discs (37)—
      No. used . . . . 3 (Bronze)
      Thickness . . . . 0.112-0.118 in.
C2 Clutch
    Piston (46)—I.D. . . . . . . 3.125-3.126 in.
      O.D. . . . . 6.600-6.610 in.
    Spring (45)—Free length . . . . . . 2.8 in.
    Test . . . . 247-302 lbs. at 1.12 in.
    Drive plates (40)—
      No. used—Wavy . . 4 (Notched Tab)
      Flat . . . . 1 (Not notched)
      Thickness-Wavy . . . 0.118-0.122 in.
      Flat . . . . 0.115-0.125 in.
    Friction discs (41)—
      No. used . . . . 5 (bronze)
      Thickness . . . . 0.112-0.118 in.
Pto Clutch
    Piston (53)—I.D. . . . . . . 3.125-3.126 in.
      O.D. . . . . 6.600-6.610 in.
    Spring (55)—Free length . . . . . . 2.8 in.
    Test, Pounds at
      inches . . . . 247-302 at 1.12
    Drive plates (58)—
      No. used—Wavy . . . 2 (Notched tab)
      Flat . . . . 1 (no notch)
      Thickness—Wavy . . 0.118-0.122 in.
      Flat . . . . 0.115-0.125 in.
    Friction discs (59)—
      No. used . . . . 3 (Fiber)
      Thickness . . . . 0.112-0.118 in.
      Minimum groove depth . . . 0.020 in.

4430
C1 Clutch
  Piston (32)—I.D. . . . . . .3.974-3.978 in.
    O.D. . . . . .7.377-7.387 in.
  Springs (33)—No. used 5
  Minimum free height of one
    spring . . . . .0.126 in.
  Drive plates (36)—
    No. used—Wavy . . .3 (Notched tab)
    Thickness—Wavy . . .0.088-0.092 in.
  Friction discs (37)—
    No. used . . . . .3 (bronze*)
    Thickness . . . . .0.127-0.133 in.
C2 Clutch
  Piston (46)—I.D. . . . . . .3.974-3.978 in.
    O.D. . . . . .7.377-7.387 in.
  Springs (44)—No. used 7
  Minimum free height of one
    spring . . . . .0.126 in.
  Drive plates (40)—
    No. used—Wavy . . .5 (Notched tab)
    Thickness—Wavy . . .0.088-0.092 in.
  Friction discs (41)—
    No. used . . . . .5 (bronze*)
    Thickness . . . . .0.127-0.133 in.
Pto Clutch
  Piston (53)—I.D. . . . . . .3.974-3.978 in.
    O.D. . . . . .7.377-7.387 in.
  Spring (54)—No. used 5
  Minimum free height of one
    spring . . . . .0.126 in.
  Drive plates (58) Before Serial
  No. T44989—
    No. used—Wavy . . .4 (Notched tab)
    Thickness—Wavy . . .0.088-0.092 in.
  Drive plates (58) After Serial
  No. T44988—
    No. used—Wavy . . .3 (Notched tab)
    Thickness—Wavy . . .0.088-0.092 in.
  Friction discs (59) Before Serial
  No. T44989—
    No. used . . . . .4 (Bronze/yellow)
    Thickness . . . . .0.127-0.133 in.
  Friction discs (59) After Serial
  No. T44988—
    No. used . . . . .3 (Fiber)
    Thickness . . . . .0.127-0.133 in.

*Bronze discs marked with white stripe are used in some applications and indicate harder friction material than similar discs marked by yellow stripe.

4630
C1 Clutch
  Piston (32)—I.D. . . . . . .3.974-3.978 in.
    O.D. . . . . .7.377-7.387 in.
  Springs (33)—No. used 5
  Minimum free height of one
    spring . . . . .0.126 in.
  Drive plates (36)—
    No. used—Wavy . . .3 (Notched tab)
    Flat . . . . .1 (Not notched)
    Thickness—Wavy . . .0.118-0.122 in.
    Flat . . . . .0.118-0.122 in.
  Friction discs (37)—
    No. used . . . . .4 (bronze*)
    Thickness . . . . .0.127-0.133 in.
C2 Clutch

Piston (46)—I.D. . . . . . .3.974-3.978 in.
  O.D. . . . . .7.377-7.387 in.
Springs (44)—No. used 7
  Minimum free height of one
    spring . . . . .0.126 in.
Drive plates (40)—
  No. used—Wavy . . .5 (Notched tab)
  Flat . . . . .1 (Not notched)
  Thickness—Wavy . . .0.118-0.122 in.
  Flat . . . . .0.118-0.122 in.
Friction Discs (41)—
  No. used . . . . .6 (bronze*)
  Thickness . . . . .0.127-0.133 in.
Pto Clutch
Piston (53)—I.D. . . . . . .3.974-3.978 in.
  O.D. . . . . .7.377-7.387 in.
Spring (54)—No. used 5
  Minimum free height of one
    spring . . . . .0.126 in.
Drive plates (58) Before Serial
No. T16988—
  No. used—Wavy . . .3 (Notched tab)
  Flat . . . . .1 (Not notched)
  Thickness—Wavy . . .0.088-0.092 in.
  Flat . . . . .0.118-0.112 in.
Drive plates (58) After Serial
No. T16987—
  No. used—Wavy . . .3 (Notched tab)
  Thickness—Wavy . . .0.088-0.092 in.
Friction discs (59) Before Serial
No. T16988—
  No. used . . . . .4 (Bronze)
  Thickness . . . . .0.127-0.133 in.
Friction discs (59) After Serial
No. T16987—
  No. used . . . . .3 (Fiber)
  Thickness . . . . .0.127-0.133 in.

*Bronze discs marked with white stripe are used in some applications and indicate harder friction material than similar discs marked by yellow stripe.

Coil type springs (45 & 55—Fig. 222) or Belleville springs (33, 44 & 54) are used to return pistons (32, 46 & 53) when hydraulic pressure is released. A suitable fixture is necessary to compress the springs for removal or installation of snap rings (34, 43 & 56). Refer to Fig. 223 for typical compressing fixture. Pistons can be removed and new seals (30, 31, 47, 48, 51 & 52) can be installed after snap rings and springs are removed. Refer to Fig. 224 for correct back to back assembly of Belleville washers. Be sure to use correct compressing fixture to prevent damage to springs or other parts while assembling.

Inside diameter of installed bushing (49—Fig. 222) should be 2.002-2.004 inches for 4230 model; 2.252-2.254 inches for all other models. Refer to paragraph 214 for inspection and service procedures for transmission pump, manifold plate and input shaft (items 8 through 26).

Refer to previously listed specifica-

tions for number of type of clutch plates and friction discs used. Most clutch packs use one flat plate and remaining plates (36, 40 & 58) are wavy. Plates with notch in one of the drive lugs (tabs) are wavy. Flat plates are not marked with notch. If a flat plate is used, the one flat plate should be assembled next to piston (32, 46 or 53). Alternate friction discs (37, 41 & 59) with plates (36, 40 & 58). Notches in tabs of wavy plates should be staggered (not aligned) and curve of the wavy plates should be alternated (not nested).

Tighten screws (61) to 20 ft.-lbs. torque. Screws which attach clutch assembly to clutch housing should be tightened to 50 ft.-lbs.; also, notice that two of these screws are shorter than remaining screws.

214. TRANSMISSION PUMP, MANIFOLD PLATE AND INPUT SHAFT. To overhaul the transmission pump, manifold plate and input shaft, first disassemble clutch pack as outlined in paragraph 213. After C1 clutch drum (28—Fig. 222) has been removed, remove the screws that hold manifold (18) to pump body (8), then slide input shaft and manifold plate assembly out of pump housing. Remove and save the

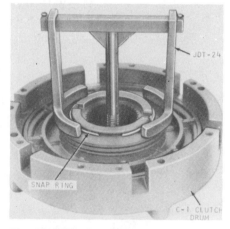

Fig. 223—Clutch 1 drum with piston and springs installed. Refer to paragraph 213 for disassembly procedure.

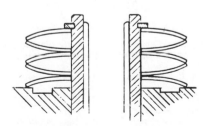

Fig. 224—Cross-sectional schematic view showing correct method of installing five Belleville washers in clutch drum. Regardless of number used (4, 5 or 7), the inner diameter of the outer washer must always fit against snap ring as shown.

two steel check balls (Fig. 225) as pump and manifold units are separated.

Remove the one remaining cap screw in rear face of pump housing (12—Fig. 222) and lift off pump body and gears.

Input shaft and bearing assembly can be removed from manifold plate after unseating and removing the retaining ring (24) from rear of manifold plate. Press bearing from shaft if renewal is

Fig. 225—When installing assembled pump on gasket and manifold, make sure steel check balls are in place as shown.

required, after removing bearing snap ring (21). Chamfer of thrust washer (23) should be toward large (rear) end of shaft (25).

Examine gears and housings for scoring, wear, cracks and other damage and renew as required. Assemble by reversing the disassembly procedure. Tighten cap screws in pump housing and body to a torque of 20 ft.-lbs. Re-assemble clutch pack as in paragraph 213.

215. **PLANETARY PACK.** Refer to paragraph 208 for removal of the planetary pack. To disassemble, place unit on bench with output end up as shown in Fig. 227. Remove the cap screws (C) and lift off C3 clutch piston housing (48). Remove the through bolts (58—Fig. 226) then lift housing (57), drum (35), housing (34) and associated parts off until planetary assembly (23) can be removed. Refer to paragraph 216 for disassembly of planetary assembly.

Refer to Fig. 226 and check parts against the values which follow:

4230

B1 Brake

Piston (6)—I.D. . . . . 9.258-9.262 in.
  O.D. . . . . 11.596-11.604 in.
Piston return plate (7)—
  Thickness . . . . 0.118-0.122 in.
Springs (8)—Free length . . . . . 0.83 in.
  Test, pounds at inch . 23-29 at 0.625
Friction discs (9)—
  Thickness . . . . 0.117-0.123 in.
  Groove depth, Minimum . . 0.020 in.
  No. used Before Serial
    No. 028121 3
  No. used After Serial
    No. 028120 2
Separator plates (10)—
  Thickness . . . . 0.118-0.122 in.
  No. used Before Serial
    No. 028121 2
  No. used After Serial
    No. 028120 1
Brake facing plate (12)—
  Thickness . . . . 0.851-0.881 in.

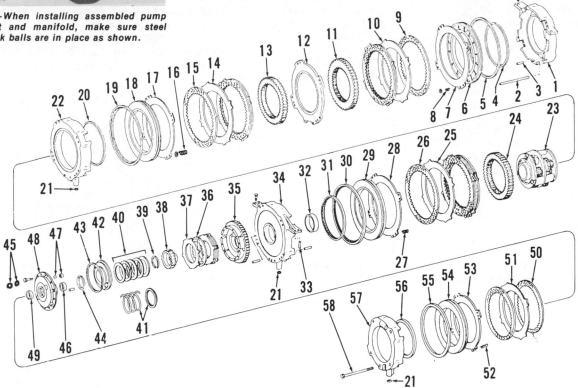

Fig. 226—Exploded view of the planetary pack. The B1 brake includes parts (4 through 11), B2 brake is (13 through 20), B3 brake is (24 through 31), C3 clutch is parts (36 through 44), B4 brake includes parts (50 through 55). Parts shown are typical and differences may be noted.

1. First planet piston housing
2. Pin (4 used)
3. Dowel pins
4. Piston inner seal
5. Piston outer seal
6. Brake piston
7. Brake piston return plate
8. Springs (used with washers on 4630 model)
9. Friction discs
10. Separator plates

11. First planet ring gear
12. Brake facing plate
13. Second planet ring gear
14. Separator plate
15. Friction discs
16. Springs (used with washers on 4630 model)
17. Brake piston return plate
18. Brake piston
19. Piston outer seal

20. Piston inner seal
21. "O" rings
22. Second planet piston housing
23. Planetary carrier assembly (Refer to Fig. 228)
24. Third planet ring gear
25. Separator plate
26. Friction discs
27. Springs
28. Brake piston return plate

29. Brake piston
30. Piston outer seal
31. Piston inner seal
32. Bushing
33. Dowel pins
34. Third planet piston housing
35. Planetary clutch drum
36. Friction discs (alternate with 37)
37. Clutch plates
38. Planetary clutch hub

39. Snap ring
40. Belleville spring washers (4630 model)
41. Coil spring and cup (used in place of 40 on 4230 & 4430 models)
42. Planetary clutch piston
43. Piston outer seal
44. Piston inner seal
45. Ball and plug
46. Bushing
47. Ball and plug

48. Clutch piston planetary housing
49. Bushing
50. Friction discs
51. Separator plate
52. Springs
53. Brake piston return plate
54. Brake piston
55. Piston outer seal
56. Piston inner seal
57. Fourth planet brake piston housing
58. Cap screws

Fig. 227—View of removed planetary pack. Refer to Fig. 226 for legend.

B2 Brake

Piston (18)—I.D......9.258-9.262 in.
　O.D............11.596-11.604 in.
Piston return plate (17)—
　Thickness..........0.118-0.122 in.
Springs (16)—Free length....0.83 in.
　Test, pounds at inch .23-29 at 0.625
Friction discs (15) Before Serial
No. 028121—
　No. used......................3
　Thickness..........0.117-0.123 in.
Friction discs (15) After Serial
No. 028120—
　No. used......................2
　Thickness..........0.117-0.133 in.
　Groove depth, Minimum ..0.020 in.
Separator plates (14)—
　Thickness..........0.118-0.122 in.
　No. used Before Serial
　　No. 028121...................2
　No. Used After Serial
　　No. 028120...................1

B3 Brake

Piston (29)—I.D......8.856-8.860 in.
　O.D............11.596-11.604 in.
Piston return plate (28)—
　Thickness..........0.118-0.122 in.
Springs (27)—Free length....1.22 in.
　Test, pounds at inch ...22-28 at 1.0
Friction discs (26)—
　Thickness..........0.117-0.123 in.
　Groove depth, Minimum ..0.020 in.
　No. used Before Serial
　　No. 028121...................4
　No. used After Serial
　　No. 028120...................3
Separator plates (25)—
　Thickness..........0.118-0.122 in.
　No. used Before Serial
　　No. 028121...................3
　No. used After Serial
　　No. 028120...................2
Bushing (32)—
　Installed diameter .3.7555-3.7570 in.

C3 Clutch

Piston (42)—I.D......3.125-3.126 in.
　O.D.............6.600-6.610 in.
Spring (41)—Free length....2.79 in.
　Test, pounds at
　　inches...........247-303 at 1.12
Drive plates (37)—
　Thickness..........0.118-0.122 in.
　No. used Before Serial
　　No. 028121...................4
　No. used After Serial
　　No. 028120...................3
Friction discs (36)—
　Thickness..........0.112-0.118 in.
　Groove depth, Minimum ..0.020 in.
　No. used Before Serial
　　No. 028121...................4
　No. used After Serial
　　No. 028120...................3
Bushing (46)—
　Installed diameter .1.8565-1.8575 in.
Bushing (49)—
　Installed diameter .1.9090-1.9100 in.

B4 Brake

Piston (54)—I.D......9.258-9.262 in.
　O.D............11.596-11.604 in.
Piston return plate (53)—
　Thickness..........0.118-0.122 in.
Springs (52)—Free length....0.83 in.
　Test, pounds at inch .23-29 at 0.625
Friction discs (50)—
　No. used......................2
　Thickness..........0.117-0.123 in.
　Groove depth, Minimum ..0.020 in.
Separator plates (51)—
　No. used......................1
　Thickness..........0.118-0.122 in.

4430

B1 Brake

Piston (6)—I.D.......8.856-8.860 in.
　O.D.............11.596-11.604 in.
Piston return plate (7)—
　Thickness..........0.118-0.122 in.
Springs (8)—Free length.....0.83 in.
　Test, pounds at inch .23-29 at 0.625
Friction discs (9)—
　Thickness..........0.117-0.123 in.
　Groove depth, Minimum ..0.020 in.
　No. used Before Serial
　　No. T44989...................3
　No. used After Serial
　　No. T44988...................2
Separator plates (10)—
　Thickness..........0.118-0.122 in.
　No. used Before Serial
　　No. T44989...................2
　No. used After Serial
　　No. T44988...................1
Brake facing plate (12)—
　Thickness..........0.307-0.317 in.

B2 Brake

Piston (18)—I.D......9.258-9.262 in.
　O.D............11.596-11.604 in.
Piston return plate (17)—
　Thickness..........0.118-0.122 in.
Springs (16)—Free length....0.83 in.

Test, pounds at inch .23-29 at 0.625
Friction discs (15)—
　Thickness..........0.117-0.123 in.
　Groove depth, Minimum ..0.020 in.
　No. used Before Serial
　　No. T44989...................3
　No. used After Serial
　　No. T44988...................2
　Groove depth, Minimum ..0.020 in.
Separator plates (14)—
　Thickness..........0.118-0.122 in.
　No. used Before Serial
　　No. T44989...................2
　No. used After Serial
　　No. T44988...................1

B3 Brake

Piston (29)—I.D......8.856-8.860 in.
　O.D............11.596-11.604 in.
Piston return plate (28)—
　Thickness..........0.118-0.122 in.
Springs (27)—Free length....1.48 in.
　Test, pounds at inch .23-29 at 1.12
Friction discs (26)—
　Thickness..........0.117-0.123 in.
　Groove depth, Minimum ..0.020 in.
　No. used Before Serial
　　No. T44989...................5
　No. used After Serial
　　No. T44988...................3
Separator plates (25)—
　Thickness..........0.118-0.122 in.
　No. used Before Serial
　　No. T44989...................4
　No. used After Serial
　　No. T44988...................2
Bushing (32)—
　Installed diameter ...3.7550-3.7575

C3 Clutch

Piston (42)—I.D.....3.1254-3.1264 in.
　O.D.............6.600-6.610 in.
Spring (41)—Free length.....2.79 in.
　Test, pounds at
　　inches...........247-303 at 1.12
Drive plates (37)—
　No. used......................4
　Thickness..........0.118-0.122 in.
　No. used Before Serial
　　No. T44989...................5
　No. used After Serial
　　No. T44988...................3
Friction discs (36)—
　Thickness..........0.112-0.118 in.
　Groove depth, Minimum ..0.020 in.
　No. used Before Serial
　　No. T44989...................5
　No. used After Serial
　　No. T44988...................3
Bushing (46)—
　Installed diameter .1.9425-1.9445 in.
　Installed depth..........3/16-in.
Bushing (49)—
　Installed diameter...2.128-2.130 in.
　Inner chamfer
　　toward..........outside (rear)

B4 Brake

Piston (54)—I.D......9.258-9.262 in.
　O.D.............11.596-11.604 in.

Piston return plate (53)—
Thickness ....0.118-0.122 in.
Springs (52)—Free length ....0.83 in.
Test, pounds at inch .23-29 at 0.625
Friction discs (50)—
No. used......................2
Thickness ....0.117-0.123 in.
Groove depth, Minimum ..0.020 in.
Separator plates (51)—
No. used......................1
Thickness ....0.118-0.122 in.

4630
B1 Brake
Piston (6)—I.D. ....9.978-9.985 in.
O.D............12.726-12.736 in.
Piston return plate (7)—
Thickness ....0.118-0.122 in.
Springs (8)—Free length .....0.83 in.
Washer used when
installing ....0.060 in.
Test, pounds at inch .23-29 at 0.625
Friction discs (9)—
Thickness ....0.127-0.133 in.
Groove depth, Minimum ..0.016 in.
No. used Before Serial
No. T16988 3
No. used After Serial
No. T16987 2
Separator plates (10)—
Thickness ....0.118-0.122 in.
No. used Before Serial
No. T16988 2
No. used After Serial
No. T16987 1
Brake facing plate (12)—
Thickness ....0.372-0.382 in.

B2 Brake
Piston (18)—I.D. .....10.155-10.161 in.
O.D............12.726-12.736 in.
Piston return plate (17)—
Thickness ....0.118-0.122 in.
Springs (16)—Free length ....0.83 in.
Washer used when
installing ....0.060 in.
Test, pounds at inch .23-29 at 0.625
Friction discs (15)—
Thickness ....0.127-0.133 in.
Groove depth, Minimum ..0.016 in.
No. used Before Serial
No. T16988 3
No. used After Serial
No. T16987 2
Separator plates (14)—
Thickness ....0.118-0.122 in.
No. used Before Serial
No. T16988 2
No. used After Serial
No. T16987 1

B3 Brake
Piston (29)—I.D. ....9.979-9.985 in.
O.D..............12.726-12.736 in.
Piston return plate (28)—
Thickness ....0.118-0.122 in.
Springs (27)—Free length ....1.22 in.

Test, pounds at inch ...22-28 at 1.0
Friction discs (26)—
Thickness ....0.127-0.123 in.
Groove depth, Minimum ..0.016 in.
No. used Before Serial
No. T16988 4
No. used After Serial
No. T16987 3
Separator plates (25)—
Thickness ....0.118-0.122 in.
No. used Before Serial
No. T16988 3
No. used After Serial
No. T16987 2
Bushing (32)—
Installed diameter .4.0675-4.0700 in.

C3 Clutch
Piston (42)—I.D. ......3.974-3.978 in.
O.D.............7.377-7.387 in.
Spring (40)—
Free height of one ....0.126 in.
No. of springs washers used 5
Drive plates (37)—
Thickness ....0.118-0.122 in.
No. used Before Serial
No. T16988 4
No. used After Serial
No. T16987 3
Friction discs (36)—
Thickness ....0.127-0.133 in.
Groove depth, Minimum ..0.020 in.
No. used Before Serial
No. T16988 4
No. used After Serial
No. T16987 3
Bushing (46)—
Installed diameter .1.9425-1.9445 in.
Installed depth ....3/16-in.
Bushing (49)—
Installed diameter ...2.128-2.130 in.
Inner chamfer
toward...........outside (rear)

B4 Brake
Piston (54)—I.D. .....10.155-10.161 in.
O.D............12.726-12.736 in.
Piston return plate (53)—
Thickness ....0.118-0.122 in.
Springs (52)—Free length ....0.83 in.
Test, pounds at inch .23-29 at 0.625
Friction discs (50)—No. used 2
Thickness..........0.127-0.133in.
Groove depth, Minimum ..0.016 in.
Separator plates (51)—
No. used..................1
Thickness ....0.118-0.122 in.

NOTE: Parts from the various tractor models and from different locations within the same tractor model may be similar, but not identical. DO NOT attempt to install similar, but incorrect parts at any location within this transmission.

Refer to paragraph 213 and Fig. 223 for suggested method of compressing spring (40 or 41—Fig. 226) in order to remove and install snap ring (39).

Bushing (32) should be pressed into housing (34) using a suitable piloted driver. Bushing should be flush with piston side of housing. Be sure to align cut-out in bushing (32) with similar cut-out in housing.

Bushing (46 & 49) should be pressed into bores of clutch housing (48) using appropriate size pilots. Front bushing (46) should be installed 3/16-inch deep in bore. Large chamfer in inside diameter of rear bushing (49) should be toward outside (rear) of clutch housing (48).

NOTE: Coat all parts with John Deere Hy-Gard transmission and hydraulic oil or equivalent while assembling.

To assemble the planetary pack, place B1 piston housing (Fig. 234) closed end down on a bench. Install four guide studs in threaded holes and alignment dowels in remaining holes as shown.

The piston return plates (7, 17, 28 & 53—Fig. 226), which are installed next to pistons, have drilled holes in the four extended lugs to serve as seats for the brake return springs. Be sure to install correct brake return springs at locations (8, 16, 27 & 52). Refer to the preceding specifications for identification at the different locations. Washers, which are 0.060 inch thick, are used with springs at locations (8 & 16) only on 4630 model.

NOTE: Be sure to align oil holes in brake facing plate (12—Fig. 226 & Fig. 227) with oil holes in B1 housing (1) when assembling. If not correctly assembled, B1 brake will not receive pressure to apply brake.

Refer to paragraph 216 for assembly of planetary assembly. Remainder of assembly procedure should be accomplished by reversing disassembly procedure and observing Fig. 226 and Fig. 227. Alternate separator plates and lined friction brake discs at all locations. All oil ports (P—Fig. 227) must be aligned as shown. Remove aligning studs, install through bolts (58—Fig. 226) and tighten to 35 ft.-lbs. torque. Tighten screws (C—Fig. 227) to 20 ft.-lbs. torque.

Before reinstalling planetary brake pack, apply 50-80 psi air pressure to each of the oil passage ports (P—Fig. 227) in turn, listen for air leaks and note action of brake plates. If leaks are noted or if brake return springs do not compress, recheck assembly procedure and correct the trouble before reassembling the tractor. Be sure the "O" rings (21—Fig. 226) in bottom of transmission case do not move when installing planetary pack. Tighten cap screws to 55 ft.-lbs. torque.

216. PLANETARY ASSEMBLY. The planetary assembly shown in Fig. 228 is located in the planetary brake pack as shown at 23—Fig. 226. Three different types of planetary units are used: Type A (Fig. 228) is typical of parts on 4230 model; Type B is typical of 4430 model; Type C is typical of planetary assembly used on 4630 model.

Refer to paragraph 208 for removal of the planetary pack and to paragraph 215 for disassembly and removal of the plantary assembly.

On type A and type B planetary assemblies shown in Fig. 228, proceed as follows: Place unit on a bench with the rear retainer (12) up. Remove three cap screws holding retainer (12) to carrier (1). Pry cover plate (11) from dowels, remove three planet pinions (10), being careful not to lose roller bearings. All rear planet pinions are equipped with **TWO** rows of 31 bearings and one spacer each. Turn planet assembly upside down and remove three cap screws holding retainer (25) to carrier. Pry cover (24) from carrier. Remove ring gear and pinions (17), being careful not to lose roller bearings.

4230 Models
Bearing rollers in pinion (6)—
    No. used each pinion ....31 x 2 rows
    Diameter of rollers ..0.0936-0.0938 in.
    Length of rollers ....0.79-0.81 in.
Thrust washer (3)
    thickness.............0.033-0.039 in.
Bearing rollers in pinion (10)—
    No. used each pinion ....31 x 2 rows
    Diameter of rollers ..0.0936-0.0938 in.
    Length of rollers ....0.79-0.81 in.
Bearing rollers in pinion (17)—
    No. used each pinion ....31 x 2 rows
    Diameter of rollers ..0.0936-0.0938 in.
    Length of rollers ....0.79-0.81 in.
Bushing (23)—
    Installed diameter.....2.130-2.132 in.
    Installed depth .....flush to 0.020 in.
                  below flush
    Notched end toward ..bottom of bore

4430 Models
Bearing rollers in pinion (6)—
    No. used each pinion ....31 x 2 rows
    Diameter of rollers ..0.0936-0.0938 in.
    Length of rollers ......0.790-0.810 in.
Spacer (5) length.......0.847-0.857 in.
Thrust washer (3)
    thickness.............0.033-0.039 in.
Bearing rollers in pinion (10)—
    No. used each pinion ....31 x 2 rows
    Diameter of rollers ..0.0936-0.0938 in.
    Length of rollers ......0.790-0.810 in.
Bearing rollers in pinion (17)—
    No. used each pinion ....24 x 2 rows
    Diameter of rollers ...0.1248-0.1250 in.
    Length of rollers ....0.790-0.810 in.
Spacer (16) length.......0.299-0.309 in.

Bushing (23)—
    Installed diameter.....2.366-2.368 in.
    Installed depth .....flush to 0.020 in.
                  below flush
    Notched end toward ..bottom of bore

4630 Models
Bearing rollers in pinion (6)—
    No. used each pinion ....20 x 2 rows
    Diameter of rollers ....0.1556 in.
    Length of rollers ....0.98-1.00 in.
Spacer (5) length.......0.330-0.340 in.
Thrust washer (3)
    thickness.............0.033-0.039 in.
Bearing rollers in pinion (10)—
    No. used each pinion ....20 x 2 rows
    Diameter of rollers ....0.1556 in.
    Length of rollers ....0.98-1.00 in.
Bearing rollers in pinion (17)—
    No. used each pinion ....20 x 2 rows
    Diameter of rollers ....0.1556 in.
    Length of rollers ....0.98-1.00 in.
Spacer (16) length.......0.094-0.104 in.
Bushing (23)—

Installed diameter.....2.505-2.507 in.
Installed depth .....flush to 0.020 in.
                   below flush
Notched end toward ...inside of bore

Model 4230 uses planetary typical of type "A" shown in Fig. 228. When assembling, refer to Fig. 230. Install the three planet pinion shafts (18) in plate (24) and locate using the steel balls (15—Fig. 228). Position retainer (25) under cover plate to hold shafts in place. Position C2-B1 sun gear (22) and the B1 ring gear (11—Fig. 226) on plate (24—Fig. 230). Locate timing marks on ring gear at center of planet pinion shafts. Install the three planet pinions (17) with bearing rollers, thrust washers and spacers and with timing marks aligned with marks on C1-B2 sun gear (21). Position thrust washer (20—Fig. 228) on sun gear, then position carrier (1) over the planet assembly being sure that dowel pins (19) properly engage

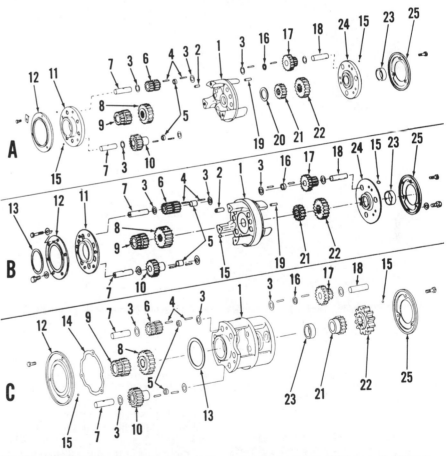

Fig. 228 – Exploded view of the three different planetary types used. Type "A" is typical of 4230; Type "B" is typical of 4430; Type "C" is typical of 4630 model.

1. Planet pinion and carrier housing
2. Dowel pins
3. Thrust washers (18 used)
4. Needle rollers
5. Spacer (6 used)
6. Third planet pinion (3 used)
7. Shafts (6 used)
8. Third planet sun gear
9. Fourth planet sun gear
10. Fourth planet pinion (3 used)
11. Pinion carrier rear cover (Types "A" & "B")
12. Pinion shaft rear retainer
13. Thrust washer (Type "B")
14. Gasket (Type "C")
15. Steel balls (9 used)
16. Spacer (3 used)
17. First and second planet pinion (3 used)
18. Pinion shaft (3 used)
19. Dowel pin
20. Thrust washer
21. Second planet (C1-B2) sun gear
22. First planet (C2-B1) sun gear
23. Bushing
24. Pinion carrier front cover (Type "A" & "B")
25. Retainer

Fig. 229—View of B1 piston housing with piston withdrawn showing seal rings. Refer to Fig. 228 for legend.

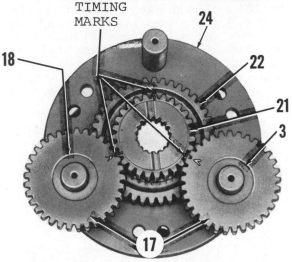

Fig. 230—View of type "A" planetary assembly showing C1 and C2 sun gears and related timing marks. Refer to Fig. 228 for legend.

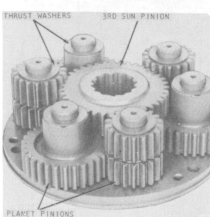

Fig. 231—Assemble planet pinion shafts, steel balls and planet pinions in rear cover plate as shown before installing sun pinion gears on type "A" unit.

and 25). Shaft is a slip fit in carrier housing bores. Each planet pinion contains two rows of twenty loose needle rollers (4) separated by a spacer for a total of forty bearings to each pinion. Withdraw pinions carefully to prevent loss of bearings. All planet pinion bearing rollers are interchangeable, but should be kept in sets. Front and rear bearing spacers (5 & 16) are of different thickness, the thicker spacers being used in rear (output) planetary unit. Notice that the first and second planet pinion (17) is index-marked "1", "2" and "3" and that corresponding marks appear on planet carrier adjacent to pinion shaft bore. When assembling planet carrier place carrier on a bench front down, install first sun gear (22) with cupped out side down and install second sun gear (21) with gear side down. Special loading tools (JDT-22-1 and JDT-22-2) can be used to more easily load the roller bearings in pinion gears. When inserting loaded pinion gears, the corresponding index marks on pinion and carrier MUST be aligned as shown in Fig. 233. No indexing is required on other planetary unit, and third sun gear (8—Fig. 228) can be installed with either side up.

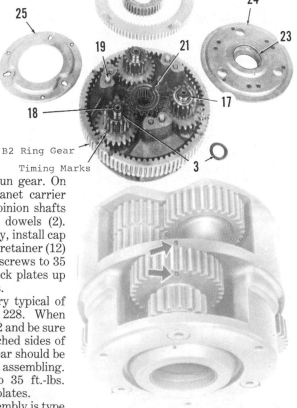

Fig. 232—View of front planet assembly of type "B" (4430) models showing ring gears and timing marks. Refer to Fig. 228 for legend.

Fig. 233—Numbers on pinion gears must align with the corresponding number stamped in housing.

correctly centered on B3 sun gear. On all models, position the planet carrier assembly (1–Fig. 228) on pinion shafts correctly engaging hollow dowels (2). Turn assembly over carefully, install cap screws through lock plates, retainer (12) and plate (11). Tighten cap screws to 35 ft.-lbs. torque, then bend lock plates up around heads of cap screws.

Model 4430 uses planetary typical of type "B" shown in Fig. 228. When assembling, refer to Fig. 232 and be sure to align timing marks. Notched sides of B1 ring gear and B2 ring gear should be away from each other when assembling. Tighten the cap screws to 35 ft.-lbs. torque and lock with lock plates.

Model 4630 planetary assembly is type "C" shown in Fig. 228. Each planet pinion shaft is retained in carrier housing by a steel ball (15) and retainer plates (12

holes in plate. Carefully turn assembly over and install the three cap screws through retainer (25) and plate (24). Tighten screws to 35 ft.-lbs. and bend lock plates around heads. Install six pinion shafts (7) and steel balls (15) in cover plate (11). Place retainer (12) under cover plate to hold shafts in plate. Install planet pinions (6 & 10) and B3 sun gear (8) as shown in Fig. 231. On models so equipped, be sure that thrust washer is

REDUCTION GEARS, TOW DISCONNECT AND PARK PAWL

All Power Shift Models

217. TRANSMISSION TOP COVER AND PARK PAWL. If park pawl remains engaged even though linkage is properly adjusted and operates satisfactorily, the park pawl return spring may be unhooked or broken. The unit can be inspected and spring renewed after removing transmission top cover. Proceed as follows:

Remove operator's platform, steering support side panels and interfering hydraulic tubes and linkage. Unbolt and remove transmission top cover. Remove damaged or broken spring and install a new spring. Malfunction can also occur if engagement spring arm (9—Fig. 243) is damaged or broken or if camshaft (5) binds. Pivot shaft (4) on some models is retained by a vertical dowel in transmission case.

218. TOW DISCONNECT. To remove the tow disconnect mechanism, first remove planetary output shaft as outlined in paragraph 208. Unscrew retaining nut (5—Fig. 244) and right hand shaft (10) and lift out disconnect fork (7), collar (4—Fig. 246) reduction gear (3) and cone of bearing (2). Front bearing cup can be drifted rearward out of housing after gear has been removed.

Install by reversing the removal procedure. Adjust reduction gear shaft bearings as outlined in paragraph 208. After tow disconnect parts have been installed, shift disconnect lever into forwardmost position with spring unhooked (Fig. 245). With cap screws loose, align the mark on stop plate with REAR of lever as shown in Fig. 245 for all models except 4630. Adjustment is similar on 4630 except mark is in **FRONT** of lever when correctly set. On all models, tighten cap screws while holding disconnect lever fully forward and mark aligned. Attach spring, and shift lever to detent (engaged) position.

219. IDLER GEAR AND SHAFT (4630 MODEL ONLY). Idler gear (17—Fig. 246) can be removed after removing planetary unit as outlined in paragraph 208 and tow disconnect and reduction gear as in paragraph 218.

Remove cotter pin, nut and washer from rear of idler shaft and snap ring (14) from front bearing; then drift or pull shaft forward out of housing and gear. Front end of shaft contains a 5/8-11 tapped hole to facilitate removal. Remove gear (17) and spacer (18) through top opening as shaft is removed. Assemble by reversing the disassembly procedure, tighten shaft nut to 180 ft.-lbs. torque and install cotter pin.

220. BEVEL PINION SHAFT. Bevel pinion shaft and bearings can be removed for service after removing planetary unit as outlined in paragraph 208 and differential assembly as in

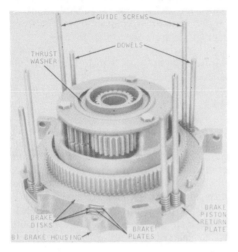

Fig. 234—Install guide screws and dowels in place of through-bolts.

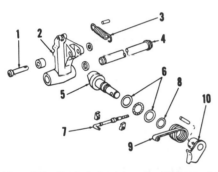

Fig. 243—Exploded view of 4630 model transmission park lock pawl and associated parts. Other models are similar.

1. Pin	6. Thrust bearing
2. Lock pawl	7. Operating cable
3. Spring	8. "O" ring
4. Pivot shaft	9. Spring arm
5. Lock cam	10. Arm hub

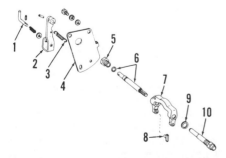

Fig. 244—View of 4230 and 4430 model tow disconnect lever. Other models are similar.

1. Latch	6. L.H. shaft & "O"
2. Shift lever	ring
3. Spring	7. Shift yoke
4. Stop plate	8. Shoe (2)
5. Retainer	9. Aluminum washer
	10. R.H. shaft

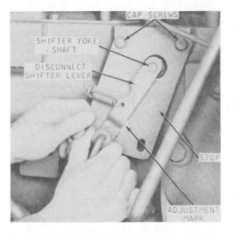

Fig. 245—Method of adjusting tow disconnect lever. See paragraph 218 for adjustment procedure.

Fig. 246—Exploded view of output gear reduction unit, main drive bevel pinion and associated parts used on 4630 model. Items 13 through 23 and item 8 are used only on 4630 model.

1. Snap ring	
2. Bearing	
3. Reduction gear	
4. Shift collar	
5. Sealing rings	
6. Oil seal	
7. Reduction shaft	
8. Drive sleeve	
9. Rear bearing	
10. Sealing bearing	
11. Sealing bearing	
12. Retainer	
13. Snap ring	
14. Snap ring	
15. Bearing	
16. Idler shaft	
17. Idler gear	
18. Spacer	
19. Snap ring	
20. Inner race	

21. Bearing	27. Spacer
22. Washer	28. Gear
23. Nut	29. Spacer
24. Nut	30. Shim pack
25. Bearing	31. Bearing
26. Shim pack	32. Pinion

paragraph 221. Remove idler shaft and gear by following procedures outlined in paragraph 219.

Remove the nut (24—Fig. 246) from front end of bevel pinion shaft and drift the shaft rearward until front bearing cone, shims (26) and spacer (27) can be removed. Withdraw shaft and rear bearing cone from rear while lifting gear (28) and spacer (29) out top opening.

The bevel pinion shaft is available only as a matched set with bevel ring gear. If rear bearing cup is renewed, keep cone point adjusting shim pack (30) intact and reinstall same pack or shims of equal thickness. If gears and/or housing are renewed, check and adjust cone point as outlined in paragraph 221.

When reinstalling bevel pinion shaft and bearings, make a trial installation using the removed shim pack (26) plus one additional 0.010 inch shim. Tighten shaft nut, then measure shaft end play using a dial indicator. Remove shims equal to measured end play plus 0.005

inch, to obtain the recommended 0.004-0.006 inch bearing preload. A preferred method of measuring bearing preload is by wrapping string around pinion shaft just in front of pinion and measuring rolling torque of shaft with a spring scale. The spring scale should indicate 2½-7½ pounds of pull required to rotate shaft one revolution per second with nut (24) tightened to 400 ft.-lbs. torque.

NOTE: If main drive bevel pinion and/or transmission housing are renewed, cone point (mesh position) of gears must be checked and adjusted BEFORE adjusting bearing preload.

Tighten nut (24) to 400 ft.-lbs. torque, then stake in at least two locations to maintain setting.

221. CONE POINT ADJUSTMENT. The cone point (mesh position) of the main drive bevel gear and pinion is adjusted by means of shims (30—Fig. 246) which are available in 0.003, 0.005 and 0.010 inch thicknesses. The cone point will only need to be checked if

the transmission housing or ring gear and pinion assembly are renewed. To make the adjustment, proceed as follows:

The correct cone point of housing and pinion are factory determined and assembly numbers are etched on rear face of pinion on all models. To determine correct thickness of shim pack (30—Fig. 246), subtract the number etched on end of pinion from the correct guide number which follows:

Model	Guide Number
4230	8.329
4430	8.329
4630	9.536

The result is the correct thickness of shim pack (30). As an example, if the guide number is 8.329 and the number etched on rear of pinion is 8.314, (8.329-8.314 = 0.015) the correct thickness of shim pack is 0.015 inch.

Adjust pinion shaft bearings to 0.004-0.006 inch pre-load AFTER cone point is correctly set. Refer to paragraph 220.

DIFFERENTIAL AND MAIN DRIVE BEVEL RING GEAR

REMOVE AND REINSTALL

All Models

222. To remove the differential assembly, first drain transmission and hydraulic fluid. Remove rockshaft housing as outlined in paragraph 282, and on all models except Power Shift, remove hex shaft as outlined in paragraph 166. Remove load control arm on all models except Power Shift. The load control arm comes out with differential on these models. On Syncro-Range and Quad-Range models, remove transmission oil pump inlet and outlet lines.

Block up tractor and remove both final drive units as outlined in paragraph 230. Remove brake backing plates and brake discs, and withdraw both differential output shafts.

If two of the ring gear to housing cap screws are recessed, rotate ring gear until the two screws are horizontal, which will provide more clearance for removal.

Place a chain around differential housing as close to bevel ring gear as possible, attach a hoist and lift the differential enough to relieve the weight on carrier bearings. Remove both bear-

ing quills using care not to lose, damage or intermix the shims located under bearing quill flanges. Differential assembly (and load control arm on power shift models) may be removed.

Overhaul the removed unit as outlined in paragraph 223.

When installing, place an additional 0.010 inch shim on bearing quill on ring gear side, tighten retaining cap screws and measure differential bearing end play using a dial indicator. Preload the carrier bearings by removing shims equal in thickness to the measured end play plus 0.002-0.005 inch. Shims are available in thicknesses of 0.003, 0.005 and 0.010 inch.

After the correct carrier bearing preload is obtained, attach a dial indicator, zero indicator button on one bevel ring gear tooth and check the backlash between bevel ring gear and pinion, in at least two places 180° apart. Proper backlash is 0.008-0.015 inch. Moving one 0.005 inch shim from one bearing quill to the other will change the backlash by about 0.010 inch.

When bearing preload and backlash are established, tighten differential bearing quill cap screws to a torque of 85 ft.-lbs. and bend up lock plates.

Tighten rockshaft housing bolts to 85 ft.-lbs. torque. Complete tractor assembly by reversing the disassembly procedure.

DIFFERENTIAL OVERHAUL

All Models

223. To overhaul the removed differential assembly, refer to Fig. 247. Most parts are interchangeable between models with or without differential lock. (Parts 7 through 14, and 19 are used with differential lock only.)

Differential gears and differential lock parts can be inspected after removing cover (6). Differential pinion shaft (18) is retained by special cap screw. Thrust washers are not used on differential pinions, but are used on axle gears. Refer to paragraph 224 for service on differential lock clutch and to paragraph 225 for bevel ring gear. Tighten cover cap screws to 55 ft.-lbs. torque if equipped with differential lock, and 85 ft.-lbs. without lock.

224. DIFFERENTIAL LOCK CLUTCH. The multiple disc differential clutch can be overhauled after remov-

ing the unit as outlined in paragraph 222 and removing cover (6—Fig. 247). The three internally splined clutch drive discs (8) are 0.112-0.118 inch thick. Discs should be renewed when thickness is less than 0.100 inch. Externally splined clutch plates (9) are 0.115-0.125 inch thick. The three piston return springs ride on guide dowels contained in cover (6) as shown in Fig. 248.

Remove piston (11—Fig. 247) with air pressure or by grasping two opposing strengthening ribs with pliers. Renew sealing "O" rings whenever piston is removed. Examine sealing surface in bore of bearing quill which houses sealing rings (19), and renew quill if sealing area is damaged. Renew cast iron sealing rings (19) if broken, scored or badly worn.

225. **BEVEL RING GEAR.** The main drive bevel ring gear and pinion are available only as a matched set. Always replace both gears. To remove ring gear, remove retaining cap screws and use a heavy drift, hammer or suitable press. Heat new gear to 300° F. in an oven and position gear. Tighten retaining cap screws to 85 ft.-lbs. torque on all models except 4630, which should be tightened to 170 ft.-lbs. Renew pinion shaft as outlined in paragraph 183 for Syncro Range models or paragraph 220 for Power Shift models.

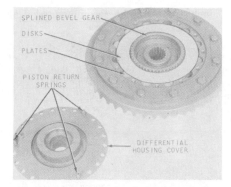

Fig. 248—Differential lock clutch partially disassembled. Piston return springs ride on guide dowels in cover and enter blank splines in clutch plate lugs.

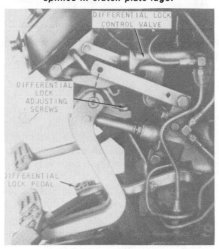

Fig. 249—With lock pedal "ON" and brake pedals "OFF" adjust screw on brake pedals to obtain approximately 0.020 clearance.

DIFFERENTIAL LOCK

Tractors may be optionally equipped with a hydraulically actuated differential lock which may be engaged to insure full power delivery to both rear wheels when traction is a problem. The differential lock consists of a foot operated control and regulating valve and a multiple disc clutch located in differential housing.

OPERATION AND ADJUSTMENT

Models So Equipped

226. Refer to Fig. 249. When pedal is depressed, pressurized fluid from the hydraulic system is directed to clutch piston (11—Fig. 247) locking axle gear (14) to differential case. Elimination of the differential as a working part causes both differential output shafts and main drive bevel gear to turn together as a unit. Available power is thus transmitted to both rear wheels equally despite variations in traction.

To release the differential lock, slightly depress either brake pedal, which releases system pressure by acting through linkage.

227. **ADJUSTMENT.** The differential lock valve is applied through an operating rod from the pedal to an over-center arm. When the pedal is depressed, the over-center feature holds valve in the applied position until either brake pedal is lightly applied. The head of lock adjusting screw then contacts a clearance bar, which is mounted crosswise on operating rod (17—Fig. 250) and pushes the rod upward to release the lock valve.

To adjust clearance, refer to Fig. 249. Depress differential lock pedal and leave pedal in the "ON" position. With brake pedals off, adjust lock adjusting screws to approximately 0.020 inch clearance between head of adjusting screw and clearance bar on operating rod. A small movement of either brake pedal will release lock valve.

Operating pressure need only be checked if valve has been overhauled or differential lock is suspected of having a malfunction. Check operating pressure by mounting a "T" fitting and a 0-1000 psi gage in the line from lock valve to differential lock. With engine running at rated speed, depress lock pedal. Gage pressure should be 420-480 psi. If it is not, disconnect links (15—Fig. 250), remove plunger (1) and add or remove shims (4). One shim should change pressure 25-30 psi.

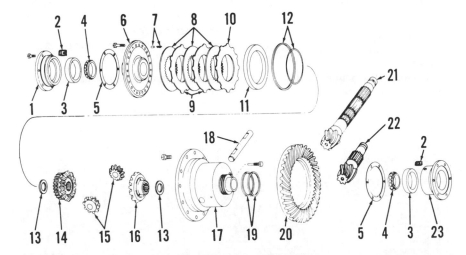

Fig. 247—Exploded view of differential assembly and associated parts. Item 2 is not used in 4630 model. Items 7 through 14 and 19 are used only in models with differential lock. Ring gear (20) is on opposite sides on some models.

1. L.H. bearing quill	7. Spring and pin (3)	13. Thrust washer	19. Sealing rings (2)
2. Square nut (2)	8. Drive disc (3)	14. Gear (diff. lock)	20. Bevel ring gear
3. Bearing cup	9. Clutch plate (2)	15. Differential pinions	21. Pinion (Syncro-mesh)
4. Bearing cone	10. Backing plate	16. Axle gear	22. Pinion (Power Shift)
5. Shim pack	11. Piston	17. Differential housing	23. R.H. bearing quill
6. Housing cover	12. "O" rings	18. Pinion shaft	

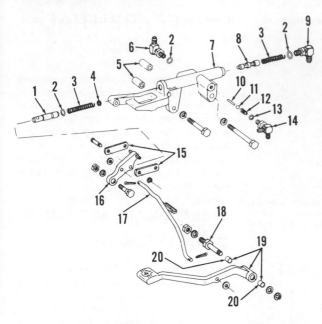

Fig. 250—Exploded view of differential lock control valve, pedal and associated parts.

1. Plunger
2. "O" ring
3. Spring
4. Steel shim
5. Spacers
6. Oil return elbow
7. Lock valve housing
8. Pressure control valve
9. Elbow (outlet)
10. Pin
11. Steel ball
12. Spring
13. "O" ring
14. Inlet "T"
15. Links
16. Arm & pin
17. Operating rod & bar
18. Special pin
19. Pedal w/bushing
20. Bushing

OVERHAUL

Model So Equipped

228. **CONTROL VALVE.** To remove the differential lock control valve, disconnect pin at plunger (1—Fig. 250), rear cap screw on brace from pedals to brake valve (as shown in Fig. 249), and the oil pipes at the valve. Remove retaining cap screws and lift off valve unit and spacer(s). Items 10 through 14 (Fig. 250) must be removed before pressure control valve (8) can be removed.

All parts are renewable individually.

Examine parts for wear or scoring and springs for distortion, and renew any parts which are questionable. Keep shim pack intact for use as a starting point when readjusting operating pressure. Renew "O" rings whenever valve is disassembled.

Assemble by reversing the disassembly procedure using Fig. 250 as a guide. Tighten control valve retaining cap screws securely and adjust as outlined in paragraph 227.

229. **CLUTCH.** Refer to paragraph 224 for overhaul procedures on differential lock clutch unit and associated parts.

on side being removed, and loosen front mount bolts on the same side for clearance on rear mount.

Support final drive assembly with a jack or a hoist, remove attaching cap screws and swing the unit away from transmission housing.

On Hi-Crop models remove the six stud nuts securing drop housing to shaft housing, remove the two retainer plugs and thread jack screws into retainer plug holes. Tighten jack screws evenly to force housings apart.

When reinstalling, be sure the sun gear (25—Fig. 251) on standard models stays all the way in, to prevent brake disc from falling behind teeth on gear. Tighten the retaining cap screws to transmission case to a torque of 170 ft.-lbs. on standard models or 150 ft.-lbs. on Hi-Crop models. Complete the installation by reversing the removal procedure. Tighten the outer gear housing (11—Fig. 252) to shaft housing (8) to a torque of 275 ft.-lbs.

OVERHAUL

All Models Except Hi-Crop

231. To disassemble the removed final drive unit, remove lock plate (24—Fig. 251) and cap screw (23), then withdraw planet carrier (21) and associated parts.

Planet pinion shaft (17) is retained in carrier by snap ring (16). To remove, expand the snap ring and, working around the carrier, tap all three shafts out while snap ring is expanded. Withdraw the parts, being careful not to lose any of the loose bearing rollers (20) in each planet pinion. Examine shaft, bearing rollers and gear bore for wear, scoring or other damage and renew as indicated.

NOTE: Bearing rollers should be renewed as a set.

REAR AXLE AND FINAL DRIVE

Model 4230 and 4430 tractors are available in high clearance (Hi-Crop) models equipped with drop housings containing a final reduction bull gear and pinion. Standard equipment on all models is a planetary reduction final drive gear located at inner ends of rear axle housings.

REMOVE AND REINSTALL

All Models

230. To remove either final drive as a unit, first drain the transmission and hydraulic fluid, suitably support rear of tractor and remove rear wheel or wheels.

On standard models, remove fenders and light wiring if so equipped. On Hi-Crop models remove 3-point hitch lift links and draft links or entire drawbar assembly. On right final drive of models so equipped, remove differential

lock pressure pipe.

On left final drive, remove rockshaft return pipe.

On models with Sound-Gard body or Roll-Gard, remove only the mount bolts

Fig. 251—Exploded view of planetary type final drive assembly typical of all models except Hi-Crop.

1. Axle
2. Oil seal
3. Retainer
4. Spacer
5. Bearing cone
6. Bearing cup
7. Lube plug
8. Axle housing
9. Dowel
10. Dowel
11. Gasket
12. Oil seal
13. Bearing cone
14. Bearing cup
15. Thrust washer
16. Retaining ring
17. Pinion shaft
18. Bearing washer
19. Planet pinion
20. Bearing roller
21. Planet carrier
22. Washer
23. Cap screw
24. Lock plate
25. Sun gear

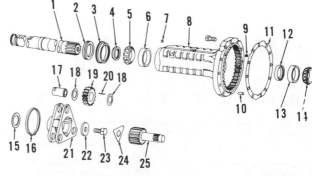

After planet carrier has been removed, axle shaft (1) can be removed from inner bearing and housing by pressing on inner end of axle shaft. Remove bearing cone (5) and spacer (4) if they are damaged or worn. When assembling, heat spacer and cone to approximately 300°F. and install on shaft making sure they are fully seated. Heat inner bearing cone (13) to 300°F. Have planet carrier ready to install. Insert bearing and install planet carrier, washer (22) and cap screw (23).

Adjustment of bearings (5, 6, 13 & 14) is accomplished by applying specified torque to screw (23), then maintaining adjustment by locking with plate (24). Correct torque should be as follows: 4030-75 ft.-lbs.; 4230-145 ft.-lbs.; 4430-175 ft.-lbs.; 4630-200 ft.-lbs.

Hi-Crop Models

232. OUTER HOUSING AND GEARS. If only outer housing, gears, shafts, bearings or oil seals are being overhauled, the complete final drive assembly will not need to be removed. Suitably support the tractor and remove wheel and tire unit. Remove draft link or disconnect drawbar from gear housing.

Remove the six stud nuts securing gear housing to shaft housing, remove the two retainer plugs and thread jack screws into retainer plug holes. Tighten jack screws evenly to force housings apart.

To disassemble the removed gear housing, first remove the bull gear cover and wheel axle inner bearing cover (16—Fig. 252 or Fig. 253), then remove inner bearing nuts (17) with a spanner wrench. Unseat snap ring on inner side of bull gear, install spacers between insides of gear housing to prevent damage to housing, then press or drive out rear axle shaft. Axle shaft is equipped with two oil seals, an inner seal which is pressed into housing and a two-piece outer seal in housing and on shaft. When assembling, heat bearing cones to a temperature of 300°F. to facilitate installation. Install and tighten inner axle shaft nut (17) to torque of 200 ft.-lbs. to seat bearing races. Rotate axle several times, then recheck torque. Loosen nut (17), then retighten to 50 ft.-lbs. Torque. On early models, install lockwasher (18) and second nut (17) to lock setting.

On later models, install locking cotter pin through axle and nut. It may be necessary to loosen nut slightly to align holes so that cotter pin can be installed. Selection of shims (24—Fig. 253) requires using a special tool which can be fabricated to dimensions shown

in Fig. 254. Assemble bearings (6—Fig. 253), gear (26), seal (28) and outer quill (9) in housing (11). Position gasket (A) over studs and install special tool over two studs as shown in Fig. 255. Tighten the two stud nuts (F) to 250

ft.-lbs. torque while making sure that screws (D) remain loose. Turn the outer drive shaft gear (26—Fig. 253) while tightening the two screws (D—Fig. 255) evenly to a torque of 25 in.-lbs. Use a feeler gage (B) as shown to

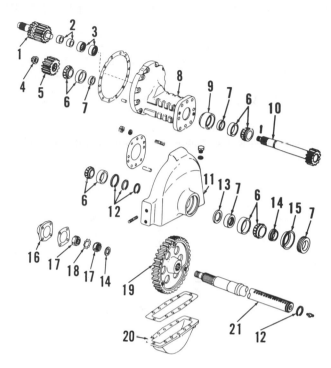

Fig. 252—Exploded view of Hi-Crop final drive assembly typical of early models. Refer to Fig. 253 for later type. Inner races (2) may or may not be used.

1. Diff. gear
2. Inner races
3. Needle bearings
4. Nut
5. Drive shaft gear
6. Bearing & cup
7. Oil seal
8. Drive shaft housing
9. Quill
10. Drive shaft
11. Gear housing
12. Snap rings
13. Washer
14. Spacer
15. Oil seal cup
16. Bearing cover
17. Bearing nut
18. Lockwasher
19. Bull gear
20. Cover
21. Axle shaft

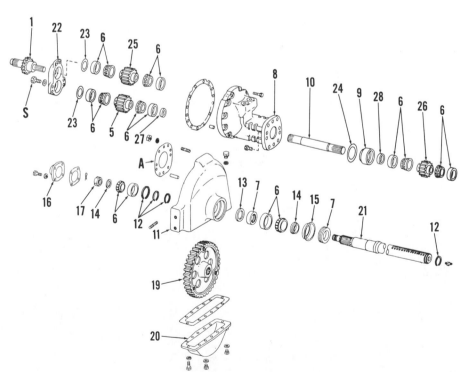

Fig. 253—Exploded view of Hi-Crop final drive assembly typical of later models. Refer to Fig. 252 for view of early type and for legend except the following.

22. Inner quill		26. Drive shaft gear
23. Shims (0.002, 0.005 & 0.010 in.)	24. Shims (0.002, 0.005 & 0.010 in.)	25. Differential output gear	27. Oil seal
			28. Oil seal

measure clearance between outer quill (C) and special tool (E). Remove special tool and install shims (24—Fig. 253) equal to measured clearance.

On early models, tighten stud nuts which attach drop housing (11—Fig. 252) to drive shaft housing (8) to 275 ft.-lbs. torque. On later models, tighten stud nuts which attach drop housing (11—Fig. 253) to drive shaft housing (8) to 445 ft.-lbs. torque. Fill each final drive housing with correct amount (3½ pints for early models; 4½ pints for later models) of SAE90 multi-purpose gear lubricant.

233. SHAFT HOUSING AND INNER GEARS. To overhaul the drive shaft housing and associated parts, first remove final drive assembly as outlined in paragraph 230 and outer gear housing as in paragraph 232.

On early models (Fig. 252), remove cotter pin and slotted nut from inner end of drive shaft (10), then remove shaft using a brass drift. When reinstalling, heat bearing cones to a temperature of 300° F. for easy installation and tighten inner nut to provide end play of 0.004 inch for shaft bearings. Drive shaft inner oil seal may be renewed when shaft is out.

On later models (Fig. 253), disassembly will be obvious after removing inner quill (22). When assembling, select thickness of shims (23) that will provide gears (5 & 25) with 0 to 0.004 inch end play. Be sure that bearings are firmly in place and that screws (S) are

tightened to 55 ft.-lbs. torque, but **DO NOT** preload bearings by installing too many shims (23). Refer to paragraph 232 for remainder of repair and adjustment procedures. Tighten cap screws which attach drive shaft housing (8) to transmission housing to 150 ft.-lbs. torque.

BRAKES

The hydraulically actuated power disc brakes use the main hydraulic system as the power source. Discs are located on differential output shafts.

OPERATION AND ADJUSTMENT

All Models

244. Power to the wet type single disc brakes is supplied by the system hydraulic pump through foot operated control valves when engine is running. A nitrogen filled accumulator provides standby hydraulic pressure in sufficient volume to apply the brakes several times after main hydraulic system ceases to operate. Control valve also contains master cylinders to permit manual operation when hydraulic pressure is not available.

Refer to Fig. 263. Parts 1 through 8 are duplicated for right and left brakes. In the first ¾-inch of pedal travel, operating rod (1) moves operating rod guide (7) which mechanically opens an equalizing valve pin and ball (1 and 6—Fig. 264). This insures equal pressure to both brakes when both pedals are depressed. Further movement of operating rod guide closes brake valve plunger (10) to close escape passage. Brake valve (12) is also unseated which allows pressure oil to fill the cavity under manual brake piston (9) and continue on to unseat check valve disc (20) which allows oil to reach brake cylinders. If pressure in valve should become too high, plunger (10) and brake valve (12) will move up as soon as pedal is released slightly, which allows escape passage to open and inlet passage to close stabilizing pressure.

In case of pressure failure, manual braking is accomplished as follows: Pressure on the operating rod and guide closes plunger (10) and opens brake valve (12). Since oil is trapped under manual brake piston (9) and the pressure line from pump is closed by a check ball and spring, pressure exerted by manual piston can open check valve seat (20) and apply brakes. As pedal is released, the reservoir check ball is pulled off its seat, allowing oil from brake valve reservoir to be pulled into

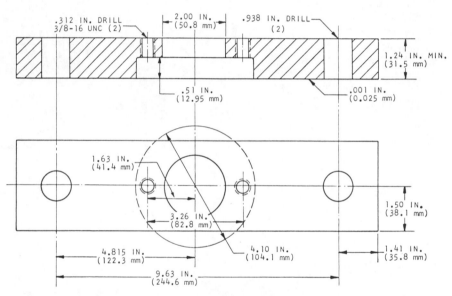

Fig. 254—The special tool shown in Fig. 255 can be fabricated to dimensions shown.

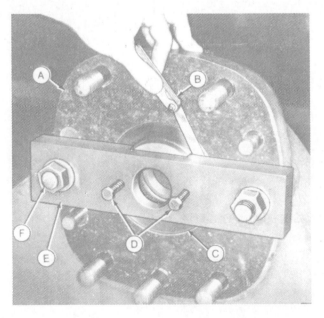

Fig. 255—Refer to text for use of special tool necessary for determining correct thickness of shims (24—Fig. 253) required for assembly.

A. Gasket
B. Feeler gage
C. Outer quill (9—Fig. 253)
D. Pressure screws
E. Special tool
F. Stud nuts

manual brake piston cavity, and continued pumping of the pedal can build pressure in the brake lines.

The only adjustment provided is at operating rod and operating rod extension (1 and 11—Fig. 263). This adjustment is used to equalize brake pedal height. To adjust the pedals, loosen locknut and turn operating rod in or out until pedals are equal in height and are approximately 5¾ inch from top of foot pad on pedal to the floor, when measured at a right angle to the floor pan (without mat). Pedal height should be equal. For service on the hydraulic pump, refer to paragraph 279. Refer to the appropriate following paragraphs for service on brake valve, actuating

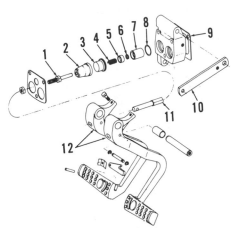

Fig. 263—Exploded view of brake valve cover, pedals and associated parts. Operating parts shown are for left brake. Right side is identical.

1. Operating rod	7. Operating rod guide
2. Boot	8. "O" ring
3. Boot retainer	9. Cover
4. Spring	10. Strap
5. Stop	11. Extension
6. Snap ring	12. Pedals

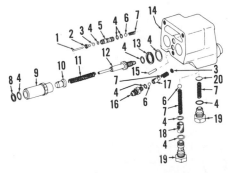

Fig. 264—Exploded view of brake valve with cover removed. Operating parts shown are for right brake.

1. Equalizing valve shaft	11. Spring
2. Sleeve	12. Inlet valve nipples (Brake valve)
3. Washer	13. Back-up ring
4. "O" ring	14. Housing
5. Guide	15. Equalizing pin
6. Steel ball	16. Inlet connector
7. Spring	17. Inlet guide
8. Paper washer	18. Screen
9. Manual brake piston	19. Plugs
10. Plunger	20. Check valve disc

cylinders, brake discs and accumulator.

BLEEDING

All Models

245. When brake system has been disconnected or disassembled, bleed the system as follows:

Pump brakes until accumulator is discharged. Start the engine, loosen lock nut on bleed screws, located on both sides of transmission housing, just above axle housings. Back out bleed screws two full turns, then tighten locknuts to prevent external oil leak. Fully depress brake pedal and hold down for 1-2 minutes. With pedal depressed, tighten bleed screw then release the pedal.

To test brake system, stop engine, discharge accumulator by holding pedal down for at least 1 minute, then depress pedal three times. Pedal travel should not exceed 4.375 inches on the third application and feel should be solid. If pedal is spongy, repeat bleeding procedure and, if trouble is not corrected, overhaul system components.

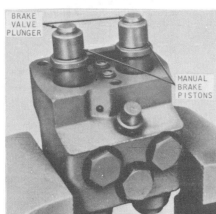

Fig. 265—Manual brake pistons and brake valve plungers can be lifted out after cover is removed.

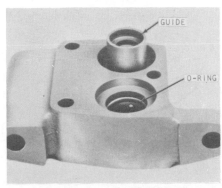

Fig. 266—Remove guides and "O" rings from cover bore.

BRAKE VALVE

All Models

246. To remove the brake valve, first bleed down the pressure in accumulator, then disconnect pressure and discharge lines at valve housing. Remove special pins in pedals which actuate operating rod extensions and remove extensions through holes in pedals. Remove strap to pedals and remove right pedal. Remove cap screws and lift off control valve.

Use Figs. 263, 264 and 265 as a guide when disassembling the brake valve. Manual brake pistons, brake valve plungers and springs can be lifted out. Use a deep socket to remove inlet valve nipples. Clean the parts thoroughly and check against the values which follow:

Brake manual piston
bore . . . . 0.9365-0.9375 in.
Manual piston O.D. . . . . . . 0.933-0.935 in.
Plunger bore in piston . . . 0.561-0.563 in.
Plunger O.D. . . . . 0.5595-0.5605 in.

Check for free movement of operating valve guide (Fig. 266) and equalizing valve pins (Fig. 267). Binding can cause equalizing valve to fail to close and both brakes will be applied if opposite pedal is pushed. Renew "O" rings whenever unit is disassembled.

Use Figs. 263, 264, 265 and 266 as a guide when reassembling. Tighten inlet valve nipples (12—Fig. 264) to a torque of 40 ft.-lbs. When control valve has been installed, check for equal pedal height and bleed the system as outlined in paragraph 245.

DISCS AND SHOES

All Models

247. To remove the brake discs or operating cylinders, first remove the final drive unit as outlined in paragraph 230. Remove the output shaft, backing plate and brake disc. The three stationary shoes are riveted to the backing plate (2—Fig. 269). The three actuating pads (4) are pressed on operating pistons (5) which can be withdrawn from transmission housing bores after brake disc is removed. Facings are available in sets of three and should only be renewed as a set.

Diameter of operating pistons (5) should be 2.2495-2.2505 inches for all models except 4630. Diameter of pistons (5) for 4630 model should be 2.6245-2.6255 inches. Pistons for all models should have 0.0025-0.0065 inch diametral clearance in cylinder bores. Refer to Fig. 268 for a schematic assembled view of brake unit.

BRAKE ACCUMULATOR

All Models

248. **R&R AND OVERHAUL.** To remove the brake accumulator, bleed fluid pressure from brake system. Disconnect pressure lines for accumulator connections, then remove retaining clip (20—Fig. 270) and remove accumulator.

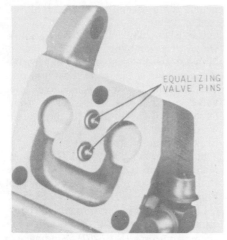

Fig. 267—Equalizing valve pins can be withdrawn from underside of cover.

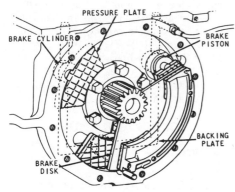

Fig. 268—Schematic view of assembled brake unit.

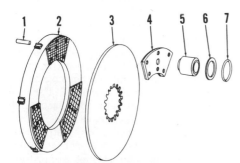

Fig. 269—Exploded view of wet type hydraulic disc brake operating parts located on differential output shafts (final drive sun gears).

1. Dowel
2. Backing plate
3. Brake disc
4. Pressure pad
5. Piston
6. Back-up ring
7. "O" ring

CAUTION: Bleed accumulator before attempting to disassemble. Gas side of accumulator piston is charged to 475-525 psi with NITROGEN gas.

Accumulator is discharged by removing protective plug (1) and depressing charging valve (9). With pressure removed, push cap (19) into cylinder, then unseat and remove snap ring (3). Remove gas end cap (4) by removing the other snap ring (3).

Check all parts for wear or damage, and assemble by reversing the disassembly procedure. Recharge the cylinder using approved charging equipment and DRY NITROGEN ONLY, to a pressure of 500 psi. Remove charging equipment and check by immersing the charged accumulator in water. When it has been determined that there are no leaks, reinstall plug (1) and install accumulator by reversing the removal procedure. Bleed brakes as outlined in paragraph 245. After engine has been run again and accumulator recharged, brakes should still have at least five power applications after engine is shut off.

POWER TAKE OFF

Tractors may be equipped with either a single speed or dual-speed pto. For ease of identification the pto types are divided into four groups, with the total number of drive gears in the system as the means of identification. The groups are called Two Gear, Four Gear, Five Gear, or Seven Gear systems. Refer to appropriate section for model being serviced.

TWO GEAR SYSTEM

All models with Syncro-Range or Quad-Range transmission and single

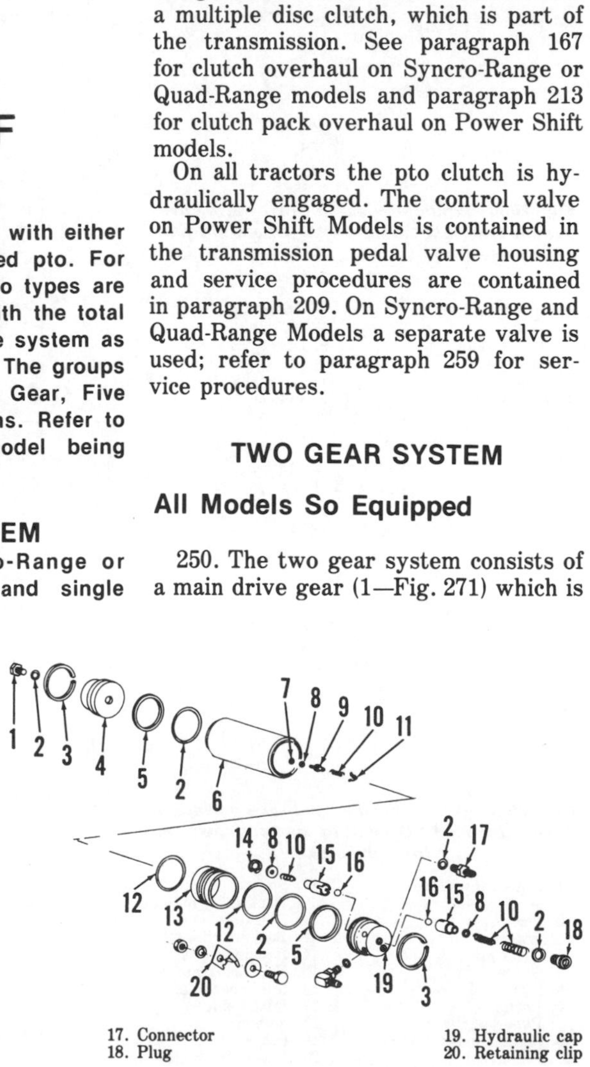

Fig. 270—Exploded view of nitrogen filled brake accumulator.

1. Plug
2. "O" ring
3. Retaining ring
4. Gas end cap
5. Back-up ring
6. Cylinder
7. Packing
8. Washer
9. Valve
10. Spring
11. Spring guide
12. Packing (U-cup)
13. Piston
14. Retaining ring
15. Guide
16. Steel ball
17. Connector
18. Plug
19. Hydraulic cap
20. Retaining clip

speed pto EXCEPT 4630 models before serial number 007000.

FOUR GEAR SYSTEM

All 4630 Power Shift models and 4630 models with Syncro-Range or Quad-Range transmission BEFORE serial number 007000.

FIVE GEAR SYSTEM

All 4030, 4230 and 4430 models with Syncro-Range or Quad-Range transmission and dual-speed pto.

SEVEN GEAR SYSTEM

All 4230 and 4430 Power Shift models with dual-speed pto.

OPERATION

All Models

249. The power take-off is driven by a hydraulically applied, independently controlled, flywheel operated, single disc clutch on Syncro-Range or Quad-Range Models. Power Shift Models use a multiple disc clutch, which is part of the transmission. See paragraph 167 for clutch overhaul on Syncro-Range or Quad-Range models and paragraph 213 for clutch pack overhaul on Power Shift models.

On all tractors the pto clutch is hydraulically engaged. The control valve on Power Shift Models is contained in the transmission pedal valve housing and service procedures are contained in paragraph 209. On Syncro-Range and Quad-Range Models a separate valve is used; refer to paragraph 259 for service procedures.

TWO GEAR SYSTEM

All Models So Equipped

250. The two gear system consists of a main drive gear (1—Fig. 271) which is

located on the clutch shaft and driven by the pto clutch hub, and the pto drive gear (7) which drives shaft (16). Bearing quill (12) and shaft can be removed as a unit, which will leave the drive gear centered on sleeve (10). Pto brake piston (8) remains applied against gear (7) when tractor is running, but pto is not in use.

251. R&R AND OVERHAUL. To overhaul the two gear system, split tractor at clutch housing as outlined in paragraph 166.

Remove clutch operating housing as outlined in paragraph 170. The large pto drive gear and oil shield must be moved rearward off sleeve (10) and then lifted out toward the front, being careful not to damage oil shield. Check bearings and renew if necessary. Check bushing (2) in main drive gear and renew if worn. Reassemble in reverse order of disassembly, and renew seal (14). Tighten rear bearing quill cap screws to 85 ft.-lbs. torque.

FOUR GEAR SYSTEM

4630 Models So Equipped

252. The four gear system is used only on 4630 Power Shift models and on other 4630 models BEFORE serial number 007000. The drive train consists of a main drive gear which is mounted behind the clutch pack on Power Shift models and is driven by the rear clutches. On other models, the main drive gear (4—Fig. 272) is mounted on the clutch shaft, and is driven by the flywheel mounted pto clutch. On all models, the main drive gear drives an upper idler (13), a lower idler (16) which is held by the pto brake (20) when not in use, and the pto drive gear (22) which drives shaft (25).

253. R&R AND OVERHAUL. The tractor must be split at the clutch housing as outlined in paragraph 170 or 204 in order to gain access to pto drive gears. On Power Shift models, remove clutch pack as outlined in paragraph 206, remove cap screws that go into transmission housing from the front, and remove clutch housing. On Quad-Range models, remove the 2-speed planetary unit as outlined in paragraph 193. On models other than Power Shift, remove transmission top cover to allow removal of the two cap screws that are screwed into rear of clutch housing and remove housing. Support the rear section of tractor with a stand on one end and a jack on the other for stability.

Use a suitable puller to remove drive gear on lower shaft after cap screw and

washer are removed. Before removing pto shaft toward the front, remove shield from the rear of shaft, puncture the oil seal (27—Fig. 272); remove seal and remove snap ring (14) inside bearing quill (29). Remove quill, and shaft can be driven out forward.

On Power Shift models, remove upper idler (6—Fig. 273). Remove retainer (15) and remove lower idler (12) with a suitable puller. Idler shaft (8) is held by set screw (7).

On other models, remove cap screw from upper idler (13—Fig. 272) and remove gear, bearings and shaft as an assembly. If necessary to disassemble, support the gear around outer edges and press, or drive, shaft (10) from gear. Remove retainer (19) and use a gear puller to pull lower idler (16) from differential pinion shaft. If it is necessary to renew bearing cups in upper or lower idler gears, use a slide hammer to remove front cup first. Spacers and snap ring can be removed and rear cup can then be driven out.

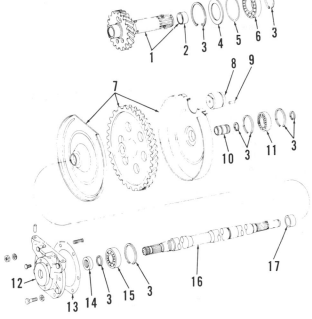

Fig. 271—Exploded view of two gear drive system used on Syncro-Range or Quad-Range tractors with single speed pto (except early 4630). "O" ring (5) is not used on 4030 and 4230.

1. Main drive
2. Bushing
3. Snap ring
4. Oil shield
5. "O" ring
6. Ball bearing
7. Pto gear w/shield
8. Brake piston
9. Dowel pin
10. Sleeve
11. Ball bearings
12. Bearing quill
13. Gasket
14. Oil seal
15. Ball bearing
16. Pto shaft
17. Bushing

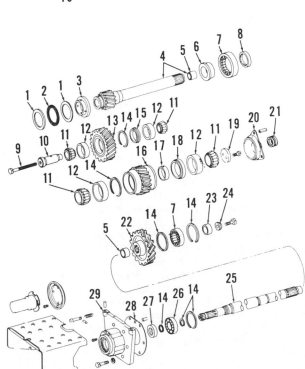

Fig. 272—Exploded view of four gear pto drive system used on 4630 Syncro or Quad-Range models before serial number 007000.

1. Race
2. Needle bearing
3. Bearing adapter
4. Pto main drive
5. Bushing
6. Inner race
7. Roller bearing
8. Sealing ring
9. Special screw
10. Idler shaft
11. Bearing cone
12. Bearing cup
13. Upper idler gear
14. Snap ring
15. Spacer
16. Lower idler gear
17. Bearing cone spacer
18. Bearing cup spacer
19. Retainer
20. Pto brake shoe
21. Brake piston
22. Pto drive gear
23. Inner race
24. Special washer
25. Pto shaft
26. Ball bearing
27. Oil seal
28. Gasket
29. Bearing quill

Renew any parts in question, and use the appropriate Fig. as a guide to reassembly. When renewing bearing cups, heat idler gears in oil (no hotter than 300°F) after installing snap ring and cup spacer. Be sure notched side of spacer is seated against snap ring. Seat bearing cups against spacer and snap ring. Heat bearing cones in oil and install on shafts while hot. Tighten cap screw (9—Fig. 272) on upper idler gear shaft to 35 ft.-lbs. torque. Install cap screws against retainer on lower idler gear and tighten against bearing cones while hot to 35 ft.-lbs. torque. Assemble in reverse order of disassembly and tighten rear bearing quill cap screws to 85 ft.-lbs. torque.

FIVE GEAR SYSTEM

All Models So Equipped

254. The five gear system is used on 4030, 4230, and 4430 models equipped with Syncro-Range or Quad-Range, and dual-speed pto. The drive train consists of a main drive gear (1—Fig. 274), pto shaft drive gear (7), pto shaft with gear (11), countershaft gear (17—Fig. 275) and 540 rpm gear (13). Pto brake piston (8—Fig. 274) is applied against gear (7) when tractor is running but pto is not in use. The 1000 rpm shaft (5—Fig. 275) goes through the 540 rpm gear and splines into pto shaft which uses only the main drive gear and pto shaft drive gear. The drive train is then like the two gear system.

When 540 rpm is desired, the 540 rpm rear shaft (2) is inserted through bearing quill and splines into 540 gear (13) which uses all five gears to transmit power to rear shaft at the reduced speed.

255. To overhaul the reduction gear assembly at rear of tractor, drain hydraulic system, remove snap ring (1—Fig. 275) from rear shaft and withdraw shaft. Remove rear bearing quill (7) and remove 540 rpm gear (13). Squeeze snap ring (12—Fig. 274) and pull pto shaft (11) from tractor, which will leave the large drive shaft gear on its sleeve (9). Countershaft gear (17—Fig. 275) can be removed by pulling shaft and snap ring (20 and 21) from case while holding countershaft gear to prevent losing the 48 roller bearings (18) and thrust washers (16). Inspect all parts for wear or other damage. When installing countershaft gear use a nonfibrous grease to hold bearings in place. Reassemble in reverse order of disassembly and tighten bearing quill cap screws and nuts to 85 ft.-lbs. torque.

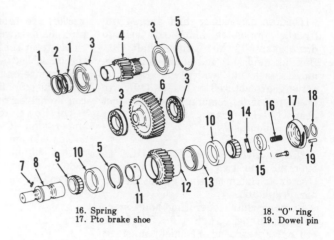

Fig. 273—Exploded view of three gears of the four gear system used on 4630 Power Shift models. Items 22 through 29 of Fig. 272 are the same on all four gear models.

1. Race
2. Thrust bearing
3. Ball bearing
4. Pto drive gear
5. Retaining ring
6. Upper idler gear
7. Set screw
8. Idler shaft
9. Bearing cone
10. Bearing cup
11. Cone spacer
12. Lower idler gear
13. Cup spacer
14. Spring plate
15. Retainer
16. Spring
17. Pto brake shoe
18. "O" ring
19. Dowel pin

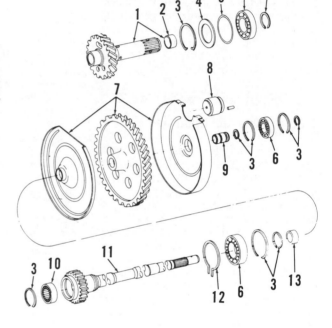

Fig. 274—Exploded view of drive gears and associated parts used in five gear system. Dual speed gears shown in Fig. 275 complete the system. "O" ring (5) is used only on 4430 models.

1. Main drive gear
2. Bushing
3. Snap ring
4. Oil shield
5. "O" ring
6. Ball bearing
7. Pto gear w/shield
8. Pto brake piston
9. Sleeve
10. Roller bearing
11. Pto shaft
12. Snap ring
13. Bushing

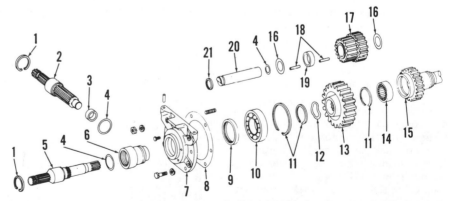

Fig. 275—Exploded view of reduction gears and associated parts used with five gear system. Item 15 is the same as Item 11—Fig. 274.

1. Snap ring
2. 540 rpm shaft
3. Bearing race
4. "O" ring
5. 1000 rpm shaft
6. Pto pilot
7. Bearing quill
8. Gasket
9. Oil seal
10. Ball bearing
11. Snap ring
12. Spring washer
13. 540 rpm gear
14. Roller bearing
15. Pto shaft
16. Thrust washer
17. Countershaft gear
18. Roller bearing (48)
19. Spacer
20. Idler shaft
21. Snap ring

If main drive gear and large pto drive gear must be removed, refer to paragraph 251 and overhaul the forward section using the same procedures as with the two gear system.

SEVEN GEAR SYSTEM

256. All 4230 and 4430 Power Shift Models with dual speed pto use this system. It consists of a four gear drive system between the planetary pack and the clutch pack in the transmission. Fig. 276 shows the main drive gear (2), upper idler gear (7), lower idler gear (10), which drives pto drive gear (12) and pto shaft (22). Pto shaft and gear (15 – Fig. 275) drives countershaft gear (17), which drives 540 rpm gear (13) at the reduced speed when 540 rpm shaft (2) is used in bearing quill (7). When the 1000 rpm shaft (5) is used, the reduction gear system is bypassed, and the four gear system in the center of transmission drives shaft (5) at 1000 rpm.

To overhaul the reduction gears, refer to paragraph 255 and Fig. 275. Pto shaft and gear (15) on this system cannot be removed from the rear unless tractor is split, since the pto drive gear (12—Fig. 276) is held to the shaft by special cap screw (14).

257. To overhaul the forward four gear drive train of the seven gear system, refer to paragraph 204 to split

tractor, and to paragraph 206 and 207 to gain access to pto gears. If only the pto main drive clutch gear (2 or 2A—Fig. 276) is to be removed, it is not necessary to remove clutch housing. Remove snap ring in front of front bearing (3 or 15) on main drive gear, remove C1 and C2 clutch shafts (Fig. 213) and pull main drive forward. If other pto gears require service, remove clutch housing as outlined in paragraph 207. Use a puller to remove pto drive gear (12—Fig. 276) after removing special cap screw and washer. Pto shaft may then be removed rearward by squeezing snap ring (23) and driving on pto shaft. Remove lower idler gear (10 or 10A), bearings and shaft assembly. On early models, upper idler gear (10) has a roller bearing on the transmission case side, but may be either a roller bearing or a ball bearing (8) on the clutch housing side. If necessary to renew ball bearing, remove with a puller. On later models, upper gear (10A) is supported in two taper roller bearings. On all models, brake piston (20) can be forced out as follows: To remove brake piston (20), remove test plug in pedal valve housing marked "PTO BR" and use air pressure in the hole to force brake piston out.

Reassemble in reverse order of disassembly. Drive pto shaft front bearing (11) into case. Heat ball bearing (24) in oil no hotter than 300° F. and install on

pto shaft after making sure snap ring (23) is installed first. Seat bearing against shoulder on shaft. If upper idler has a ball bearing, heat bearing before installing. Snap rings (26, 27 & 31) are available in several thicknesses for later models. Thickest snap ring possible to install should be used to reduce bearing play to near zero. On 4230 and 4430 models, tighten cap screw (14) to 130 ft.-lbs. torque. On later models tighten screw (14) to 120 ft.-lbs. torque. On all models, tighten rear bearing quill cap screws and nuts to 85 ft.-lbs. torque.

PTO CLUTCH VALVE

Power Shift Models

258. The Power Shift pto clutch valve service is covered in paragraph 209 as part of the pedal valve.

To adjust the pto operating linkage on Power Shift models, remove the left cowl for access to clutch valve operating rod. Disconnect yoke from upper end of rod and push pto lever to 3/16-inch from front of lever slot (front edge of lever). Push down on operating rod until valve bottoms and adjust yoke so pin hole aligns with lower hole in lever.

Syncro-Range and Quad-Range Models

The pto clutch valve is included as part of the Perma-Clutch valve and oil pressure regulating valve housings, located on the left side of clutch housing. The inner section of the housing (12—Fig. 160) contains the clutch and pto operating valves and the pto lock piston (25), which is applied hydraulically after pto valve has been engaged. This keeps pto lever engaged firmly until lever is disengaged manually. The outer section of housing (2) contains the Perma-Clutch oil pressure regulating valve and the oil cooler relief valve.

259. **REMOVE AND REINSTALL.** To remove outer and inner sections of oil pressure regulating valve housing, remove operating rods for both clutch valve and pto valve (Fig. 287). Loosen locknut on pto operating rod to allow the yoke pin retainer to be lifted out of pin. On 4630 models serial no. 007000 and later drive the retainer spring pins out of arms (Fig. 288) and remove rods. Remove oil pipes, clutch pedal return spring and remove only the eight outermost flange cap screws and pull assembly straight outwards to get free of adapters (See items 14 and 15 Fig. 160). After the outer and inner sections are removed as a unit, the inner cap screws can be removed to separate the

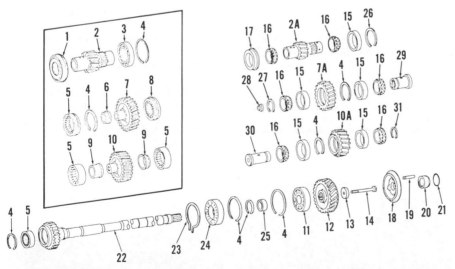

Fig. 276—Exploded view of pto drive gears of seven gear system. Reduction gear portion of drive train is shown in Fig. 275. Early type is shown in inset. Bearing shown (8) may be either roller or ball type bearing. On late models, install thickest snap ring (26, 27 and 31) possible to correctly adjust bearings.

1. Ball bearing
2 & 2A. Pto main drive
3. Ball bearing
4. Snap ring
5. Roller bearing
6. Inner race
7 & 7A. Upper idler gear
8. Ball bearing
9. Inner race
10 & 10A. Lower idler
11. Ball bearing
12. Pto drive gear
13. Washer
14. Special cap screw
15. Bearing cup
16. Bearing cone
17. Flanged bearing cup
18. Brake shoe
19. Pin
20. Brake piston
21. "O" ring
22. Pto shaft & gear
23. Snap ring
24. Ball bearing
25. Bushing
26. Snap ring
27. Snap ring
28. Shaft retainer
29. Upper idler gear shaft
30. Lower idler gear shaft
31. Snap ring

halves. Inspect the "O" rings on adapters and renew if necessary. Assemble in reverse order of disassembly.

260. To overhaul the pto clutch valve and Perma-Clutch valve in the inner housing, separate inner and outer housings (Fig. 160). The pins in valves (10 and 11) must be removed before springs and valves will come out. The pto valve, springs, pin and spacers are identical with those of the Perma-Clutch, but it is better to return the valves to the bores from which they were removed. Inspect valves for nicks or burrs and make sure valves slide freely when reinstalled. Renew "O" rings on adapters if necessary and be sure pto lock piston (25) is free in its bore. Tighten cap screws to outer housing to 20 ft.-lbs. torque.

261. The oil pressure regulating valve housing (2—Fig. 160) contains the oil pressure regulating valve (23) and oil cooler relief valve (22). Valve (23) maintains approximately 140 lbs. pressure to the power train, and opens above that pressure to allow oil to be routed to the main pump and oil cooler. Make sure valves move freely in their bores and that they are free of nicks and burrs. Late model valves have "O" rings on them to prevent sticking. Shims (19) must be in place before assembly. Refer to paragraph 162 for clutch pressure check.

HYDRAULIC SYSTEM

The closed center hydraulic system provides standby pressure to all tractor hydraulic components, with a maximum available flow of 16.5 gpm on 4030 models, 20.6 gpm on 4230, 4430 and 4630 models without Power Front Wheel Drive, 23.5 gpm on 4230, 4430 and 4630 models with Power Front Wheel Drive. Models with Power Front Wheel Drive have a 4.0 cubic inch main pump, while all other models are equipped with a 3.0 cubic inch pump. Working fluid for all hydraulic functions is available at a standby pressure of 2200-2300 psi.

MAIN HYDRAULIC SYSTEM

All Models

272. **OPERATION.** The main hydraulic system pump is mounted in front of radiator and coupled to engine crankshaft. This variable displacement, radial piston pump provides only the fluid necessary to maintain system pressure. When there are no demands on the system, pistons are held away from the pump camshaft by fluid pressure and no flow is present. When pressure is lowered in the supply system by hydraulic demand or by leakage, the stroke control valve in the pump meters fluid from the camshaft reservoir, permitting the pistons to operate and supply the flow necessary to maintain system pressure.

The transmission pump provides pressure lubrication for the transmission gears and shafts, and has a small priority valve in the pump manifold, which closes off the lubricating pressure until the clutch and main pump receive a preset pressure. On Power Shift Models, the pump also supplies the operating fluid for transmission operation. On all models, excess fluid flow from transmission pump passes through the full flow system filter to the inlet side of the main hydraulic system pump. If no fluid is demanded by the main pump, the fluid passes through the oil cooler then back to reservoir in transmission housing.

The oil cooler is mounted in front of tractor radiator and on air-conditioned models is an integral part of the air-conditioning condenser.

Return oil from the different functions is routed through a second filter on Power Shift models.

273. **RESERVOIR AND FILTERS.** The hydraulic system reservoir is the transmission housing and the same fluid provides lubrication for the transmission gears, differential and final drive units. The manufacturer recommends that only John Deere Hy-Gard transmission and Hydraulic Oil or its equivalent be used in the system. Reservoir capacity in U.S. Gallons is as follows:

Power Shift Models
4230	15½
4430	14
4630	16

Syncro-Range and Quad-Range Models
4030	14
4230	16
4430	16
4630	20

Approximately 5 U.S. Gallons must be added to above capacities on all tractors equipped with Power Front Wheel Drive.

To check the fluid level, stop tractor on level ground and check to make sure that fluid level is in "SAFE" range on dipstick (Fig. 289).

The oil filter element (or elements) may be renewed without draining fluid reservoir, by removing filter cover and extracting element. Filters are located on left side of transmission housing as shown in Fig. 290.

All filters are provided with a bypass valve which opens to allow oil to flow when cold or with filter plugged. On Syncro Range models, the by-pass valve is located in oil filter relief valve housing (Fig. 291). To service filter relief valve, remove front plug and with-

Fig. 287—View of pressure regulating valve housing and operating linkage typical of all models with Perma-Clutch except 4630.

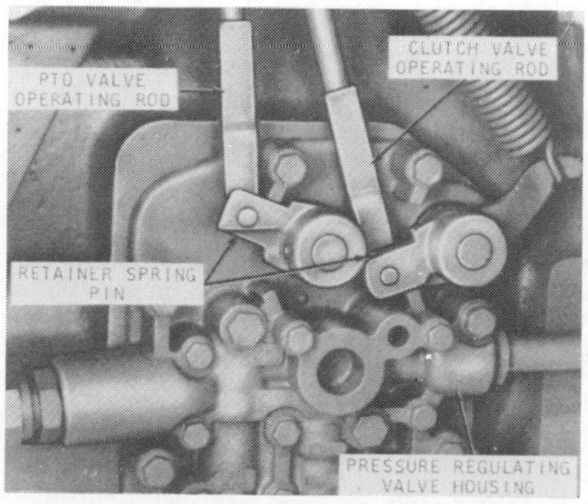

Fig. 288—View of 4630 model operating linkage on pressure regulating valve housing for tractors with Perma-Clutch.

Fig. 289—Hydraulic system and transmission fluid filler cap and dipstick are located at rear of tractor as shown.

draw spring and valve plunger. The housing also contains a return check valve, and the early model housing contains a manual by-pass valve. Late model housings had "O" rings on the valves, but some return oil check valves had a groove without an "O" ring and the manual by-pass valve was discontinued.

On Power Shift models the relief valve for front (transmission) filter is located in the Power Shift Regulating Valve housing and is shown at (11—Fig. 219). The rear (hydraulic) filter by-pass valve is located in filter base housing as shown in Fig. 292. To renew or check the valve it is necessary to drain the system and remove rear filter; then unbolt and remove filter housing.

The manual by-pass valve (22-Fig. 218 for early Power Shift Models; or Fig. 291 for other early models) applies less restriction to the return oil from a single acting cylinder, but allows a greater quantity of oil to by-pass the oil cooler on Power Shift models. Open the valve if necessary, when using a front loader or similar implement, but make sure valve is closed for normal tractor operation. By-pass valve can be turned in or out after removing the hex cap nut. On all late models, the by-pass valve was discontinued.

274. SYSTEM TESTS. Efficient operation of the tractor hydraulic units requires that each component operates properly. A logical procedure for testing the system is therefore needed. The indicated tests include Transmission Pump Flow Test, System Pressure Tests and Leakage Tests as outlined in the following three paragraphs. Unless the indicated repair of hydraulic units is obvious because of breakage, these tests should be performed before attempting to repair the individual units.

275. TRANSMISSION PUMP FLOW TEST. A quick test of transmission pump operation can be performed by removing the fluid filter (front filter on Power Shift Models) and turning engine over with starter. A generous flow of fluid will be pumped into filter housing if pump is operating satisfactorily.

To more thoroughly test pump condition, connect a flow meter into main hydraulic pump supply line on left side of clutch housing (see Fig. 293 or Fig. 294). With unit at operating temperature, engine at 1900 rpm and flow unrestricted, output should be 10 gpm for all Syncro-Range or Quad-Range Models, or 12 gpm for all Power Shift Models.

Slowly close test unit pressure valve while observing flow, which should remain relatively constant to 90 psi. At

approximately 100 psi, relief valve should start to open and flow decrease; if it does not, overhaul Oil Cooler Relief Valve (22—Fig. 160) on Syncro-Range Models or (6—Fig. 218 or Fig. 219) on Power Shift Models.

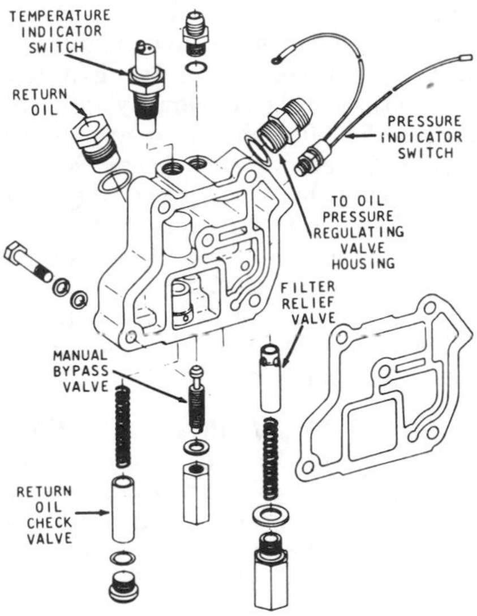

276. MAIN PUMP PRESSURE AND FLOW. To check the main pump pressure and flow, bleed off hydraulic pressure by opening the right brake bleeder screw and holding pedal down a few moments. Remove the line leading from pressure oil manifold (Fig. 295) to pressure control valve. Connect inlet side of flow meter to manifold and connect outlet line from flow meter as shown in Fig. 296 or Fig. 297 on left side of tractor. Start and run engine at 1800 rpm.

NOTE: If flow meter is not the inline type, install outlet hose in the transmission filler tube. Run engine for short periods only while testing with this type meter.

Close the test unit control valve until a pressure of 2000 psi is registered; then check fluid flow which should be 16 gpm for 4030 models; 20.6 gpm for 4230, 4430 and 4630 tractors without Power Front Wheel Drive, 23.5 gpm for 4230, 4430 and 4630 models with Front Wheel Drive.

Slowly close flow meter control valve

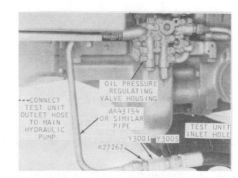

Fig. 291—*Exploded view of oil filter relief valve housing and associated parts used on early model tractors with Perma-Clutch. Late models have "O" ring grooves and "O" rings on valves, but do not have manual by-pass valve. Corresponding valves on Power Shift units are shown in Fig. 218 and Fig. 219.*

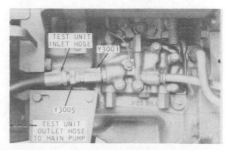

Fig. 293—*Connect flow meter as shown for transmission pump test on Syncro or Quad-Range models.*

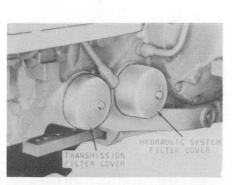

Fig. 290—*Typical Power Shift dual filter arrangement showing return function filter behind transmission filter. Other models have only a transmission filter.*

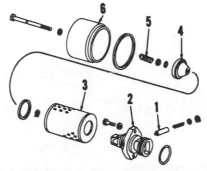

Fig. 292—*Rear hydraulic filter and associated parts used on Power Shift models.*

1. By-pass valve
2. Base housing
3. Filter element
4. Retainer
5. Spring
6. Cover

Fig. 294—*Connect flow meter as shown to test transmission pump on Power Shift models.*

until flow stops; pressure should be 2250-2300 psi for models with Power Front Wheel Drive or 2200-2300 psi for other models. Adjust standby pressure if necessary, by turning the stroke control adjusting screw (2—Fig. 299) clockwise to increase standby pressure or counter-clockwise to decrease pressure.

277. PRESSURE CONTROL (PRIORITY) VALVE CHECK. Be sure hydraulic pressure is bled off. Remove flow meter and reconnect line to pressure control (priority) valve. Refer to Fig. 295 and install a 3000 psi pressure gage in test plug hole. Install a jumper

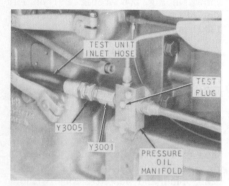

Fig. 295—Install tester inlet hose on manifold after removing line to pressure control valve.

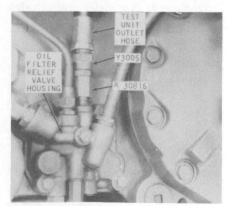

Fig. 296—Install tester outlet hose for Syncro and Quad-Range models in filter relief valve housing as shown.

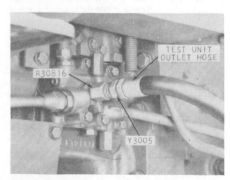

Fig. 297—Install outlet hose for Power Shift models in pressure regulating valve housing as shown.

hose in a breakaway coupler so that hydraulic fluid can flow through the coupler and back to the reservoir, and turn metering valve arm to "Fast" position.

Start engine and operate at 800 rpm (slow idle), then move selective control valve lever to pressurize breakaway coupler. Gage reading will be the minimum pressure which is maintained by priority valve to insure that steering and brakes will always be pressurized, even if other functions receive no pressure. If pressure is not within recommended range of 1650-1700 psi, refer to Fig. 19 and disassemble and inspect priority valve. BE SURE hydraulic pressure is bled off before removing any oil lines. Shims control the pressure at which control valve starts to restrict oil flow to functions.

278. LEAKAGE TEST. To check for leakage at any of the system valves or components, move all valves to neutral and run engine for a few minutes at 1900 rpm. Check all of the hydraulic unit return pipes individually for heating. If the temperature of any return pipe is appreciably higher than the rest of the lines, that valve is probably leaking. Disconnect that return line and measure the flow for a period of one minute. Leakage should not exceed ½-pint; if it does, overhaul the indicated valves as outlined in the appropriate sections of this manual.

279. MAIN HYDRAULIC PUMP. When external leaks or failure to build or maintain pressure indicates a faulty hydraulic pump, remove the unit for service as follows:

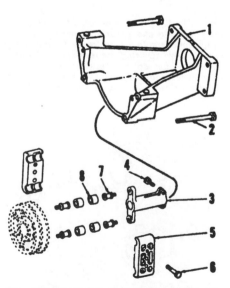

Fig. 298—Exploded view of main hydraulic pump support and drive coupling.

1. Support	5. Coupler half
2. Cap screw	6. Cap screw
3. Driveshaft	7. Coupler stud
4. Cap screw	8. Bushing

Relieve hydraulic pressure. The oil cooler and radiator must both be drained and then removed. Disconnect main supply line, oil cooler line, pressure line and oil seal drain tube from pump. Remove drive shaft coupler halves and loosen clamp screws in pump half of coupler. Suitably support the pump, remove screws securing pump to support and remove the pump.

Install by reversing the removal procedure. Tighten pump mounting bolts to a torque of 85 ft.-lbs. Other applicable torques are as follows: Note: Numbers in bolt description refer to identification symbols in Fig. 298.

Pump support to cyl. block (2) 85 ft.-lbs.
Pump drive clamp screws(4) . .35 ft.-lbs.
Drive coupler cap screws (6) .35 ft.-lbs.
Drive coupler studs (7) . . . . . .35 ft.-lbs.
Power Front Wheel Drive Keeper to
  drive shaft cap screws . . . . .20 ft.-lbs.
Use "Loctite" when installing coupler studs (7).

280. OVERHAUL. Before disassembling the pump, check pump shaft end

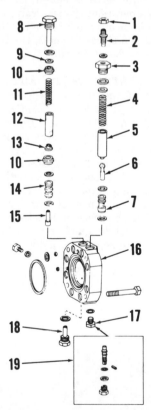

Fig. 299—Exploded view of stroke control valve housing showing component parts. Item 19 might not be used.

1. Nut	11. Spring
2. Adjusting screw	12. Filter
3. Bushing	13. Guide
4. Spring	14. Valve sleeve
5. Spring guide	15. Outlet valve
6. Stroke control valve	16. Housing
7. Valve sleeve	17. Plug
8. Plug	18. Plug
9. Shim washer	19. Lockout valve
10. Packing	

play using a dial indicator, and record the measurement for convenience in reassembling. End play should be 0.001-0.003 inch for either (3.0 or 4.0 cubic inch) pump. End play is adjusted by adding or removing shims (20—Fig. 300) which are available in thickness of 0.006 and 0.010 inch. Bearing wear, or wrong number of adjusting shims can cause excessive end play.

To disassemble the pump, remove the four cap screws retaining the stroke control valve housing (16—Fig. 299) to front pump, and remove the housing. Withdraw discharge valve plugs, guides, springs and valves (10—Fig. 300). Remove all piston plugs (1), springs (2) and pistons (3); then carefully withdraw pump shaft (15) together with bearing cones, thrust washers (14), cam race (17) and the loose needle rollers (16). Thrust washers (14) are 0.1235-0.1265 inch thick for 3.0 cubic inch pump (models without Power Front Wheel Drive); 0.0422-0.0452 inch thick for 4.0 cubic inch pump used on models with Power Front Wheel Drive. There are 36 bearing rollers (16) used on 3.0 cubic inch pumps; 25 bearing rollers (16) used on 4.0 cubic inch pumps.

Remove the plugs retaining inlet valve assemblies (5) and check inlet valve lift using a dial indicator. Lift should be 0.060-0.080 inch. If lift exceeds 0.080 inch, spring retainers are probably worn and valves should be renewed. Also check for apparent excessive looseness of valve stem in guide. Do not remove inlet valve assembly unless renewal is indicated or discharge valve seat (9) must be renewed. To remove the inlet valve, use a small pin

punch and drive valve out, working through discharge valve seat (9). If inlet valve is to be re-used, place a flat disc on inlet valve head from the inside, so that all the driving force will not strike valve head in the center. Be sure the disc will drive through the hole without touching. Discharge valve seat can be driven out after inlet valve is removed. Be sure to reinstall pistons, springs, valves and seats to their own respective bores. The piston bores in all pumps are lined with a Teflon sheath, so all bores should be carefully inspected. Scored pistons or bores could cause pistons to stick.

The manufacturer recommends that the eight piston springs (2) test within 1½ lbs. of each other and within range of 34-40 lbs. at 1.62 inches for 3.0 cubic inch pump; 47-53 lbs. at 1.78 inches for 4.0 cubic inch pump. Install seal (6) only deep enough to allow snap ring to enter groove, to avoid blocking the relief hole in body.

Valves located in stroke control valve housing control pump output as follows: The closed hydraulic system has no discharge except through the operating valves or components. Peak pressure is thus maintained for instant use. Pumping action is halted when line pressure reaches a given point by pressurizing the camshaft reservoir of pump housing, thereby holding pistons outward in their bores.

The cutoff point of pump is controlled by pressure of spring (4—Fig. 299) and can be adjusted by turning adjusting screw (2). When pressure reaches the standby setting, valve (6) opens and meters the required amount of fluid at reduced pressure into crank-

case section of pump. Crankcase outlet valve (15) is held closed by hydraulic pressure and blocks the outlet passage. When pressure drops as a result of system demands, crankcase outlet valve is opened by the pressure of spring (11) and a temporary hydraulic balance on both ends of valve, dumping the pressurized crankcase fluid and pumping action resumes. Stroke control valve spring (4) should test 125-155 lbs. pressure when compressed to 3.3 inches, and crankcase outlet spring (11) should test 45-55 lbs. at 2.2 inches.

Cutoff pressure is regulated by the setting of adjusting screw (2) and adjustment procedure is given in paragraph 276. Cut-in pressure is determined by the thickness of shim pack (9) and/or pressure of spring (11). A special tool (JDH-19) is available to determine shim pack; refer to Fig. 301 and proceed as follows:

Assemble outlet valve units (8 through 15—Fig. 299), using existing shim pack (9). Install special tool (JDH-19) in place of plug (18), using one 1/8-inch thick washer as shown in Fig. 301. If adjusting shim pack thickness is correct, scribe line on tool plunger should align with edge of tool plug bore as shown; if it does not, remove top plug (8—Fig. 299) and add or remove shim washers (9) as required. Shims (9) are available in 0.030 inch thickness. If special tool is not available use shim washers of same thickness as those removed, then add shims to raise cut-in pressure, or remove shims to lower pressure.

When installing stroke control housing, add or remove shims (20—Fig. 300) as necessary to obtain specified pump shaft end play of 0.001-0.003 inch. Always use new "O" rings, packings and seals. Oil all parts liberally with clean hydraulic system oil. Tighten stroke control valve housing retaining cap screws to 85 ft.-lbs. and tighten piston cap plugs to 100 ft.-lbs. torque. Adjust standby pressure as outlined in paragraph 276 after tractor is reassembled.

281. **PRIORITY VALVE.** The Pressure Control (Priority) valve is mounted on right side of rockshaft housing, just ahead of rockshaft. Refer to paragraph 15 or 23 in appropriate Steering Section for data on the valve unit.

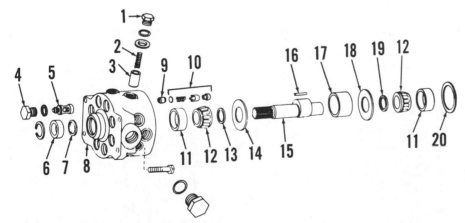

Fig. 300—Exploded view of main hydraulic pump body, shaft and associated parts. Two additional spacers are used between bearing cones and thrust washers on 4.0 cid pumps.

1. Plug	6. Oil seal
2. Spring	7. Packing
3. Piston	8. Body
4. Plug	9. Seat
5. Inlet valve	10. Discharge valve

11. Bearing cup	16. Roller bearing
12. Bearing cone	17. Race
13. Spacer	18. Thrust washer
14. Thrust washer	19. Spacer
15. Pump shaft	20. Shim

ROCKSHAFT HOUSING & COMPONENTS

All Models So Equipped

282. **REMOVE AND REINSTALL.**

To remove the rockshaft housing, first remove seat and operator's platform. Disconnect the three-point lift links on tractors so equipped. Disconnect and remove hydraulic control rods, interfering wiring and hydraulic lines. Remove the attaching bolts and lift the housing from tractor using a hoist.

When installing the unit, place load selector lever in "MAX" position and make sure linkage roller is to rear of cam follower as housing is lowered. Do not bend draft linkage. Complete the assembly by reversing the removal procedure and tighten the retaining cap screws to a torque of 85 ft.-lbs.

283. CONTROL VALVE HOUSING OVERHAUL. To remove the control valve housing, disconnect operating rods; then unbolt and lift off the cover.

NOTE: If tractor is equipped with dual or triple selective control valves, housings may be unbolted and laid aside before valve housing is removed. Be sure rockshaft valve cover remains with valve.

Remove rear cover (20—Fig. 302) (or selective valve mounting cover on models so equipped), and withdraw operating valves (16 through 19) and flow control valve (11-13). Remove thermal relief valve assembly (7) and plug (6).

Remove load selector arm (22—Fig. 303), load selector control shaft (23), selector control arm (25) and links with roller (24). Remove valve camshaft (1), and avoid losing washer as camshaft is

removed. Disconnect spring (4) from control valve adjusting cam (5), remove retainer ring holding linkage to control valve adjusting link (6) and remove linkage.

Check all linkage, springs, valves and housing for wear, scoring, or other damage and renew any parts in question. Valves can be lapped to seats, if necessary, using fine lapping compound. Inspect thermal relief valve assembly and spring, which should have 8 to 10 pounds pressure at a compressed length of 15/32-inch.

When reassembling, install thermal relief valve, spring and shims as required. Refer to Figs. 305, 306 & 307 for assembly sequence of control linkage. Install valve operating link (10—Fig. 303), link with pin (8) and spring. Assemble control valve operating cam (5) and adjusting link (6), adjusting screw and screw nut (2 and 7). Install in housing and attach to link pin. With control valve operating cam in position,

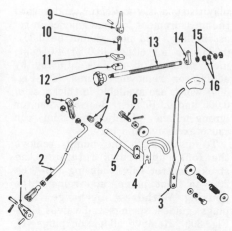

Fig. 304—Exploded view of typical rockshaft control lever and depth adjusting screw.

1. Lower operating arm	9. Cam
2. Control rod	10. Height stop screw
3. Control lever	11. Height stop
4. Friction plate	12. Special nut
5. Shaft	13. Adjusting screw
6. Friction screw	14. Lever stop
7. Bushing	15. Special washers
8. Upper operating arm	16. Spring washers

Fig. 302—Exploded view of rockshaft control valve housing, valves and associated parts. Refer to Figs. 303 and 304 for control linkage.

1. Housing	
2. Gasket	
3. "O" ring	
4. Back-up ring	
5. Inlet pipe	
6. Adjusting plug	
7. Thermal relief valve	
8. Pipe plug	
9. Plug	
10. Check ball	
11. Flow control valve	
12. Spring	
13. Shim	
14. Adjusting plug	
15. Guide	
16. Metering shaft	

17. Valve seat	19. Spring
18. Valve ball	20. Cover

Fig. 303—Exploded view of typical rockshaft control linkage. Refer to Fig. 304 for rockshaft control lever and depth adjusting screw. Item 20 is the same as item 1 in Fig. 304.

1. Valve cam shaft	
2. Adjusting screw	
3. Anchor pin	
4. Spring	
5. Valve operating cam	
6. Cam adjusting link	
7. Adjusting screw nut	
8. Link w/pin	
9. Operating link w/cam	
10. Operating link w/pins	
11. Operating shaft	
12. Eccentric	
13. Star washer	
14. "O" ring	
15. Bushing	
16. Spring	
17. Washer	
18. Quill	
19. Spring	
20. Lower operating arm	
21. Selector rod	
22. Load selector arm	
23. Shaft	
24. Links w/roller	

25. Selector control arm	28. Upper operating arm
26. Spring	29. Shaft
27. Bushing	30. Load selector lever

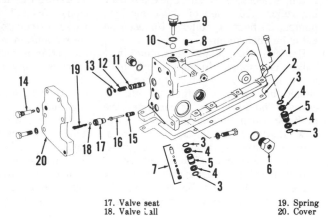

Fig. 301—Special adjusting tool (JDH 19) can be used to determine stroke control valve setting. Refer to paragraph 280.

1/8" OR .125"
SCRIBE LINE
JDH-19 CRANK-CASE OUTLET VALVE ADJUSTING TOOL
ADJUSTING WASHERS

install valve camshaft (1) and washer, and hook spring to valve operating cam tang. Install operating lever shaft quill (18) and lower operating arm (20), with shaft (11) in housing. Be sure pin on operating link is in hole of valve operating shaft (11). Assemble load selector links with roller (24) and arm (25) in housing and attach valve operating link (10) to operating link with cam (9). Install lower operating arm spring pin and install pin in load selector control arm (25). Install valve assembly in reverse order of disassembly, and make sure that the roller of the load selector control link goes on the backside of cam follower. Tighten retaining cap screws to a torque of 85 ft.-lbs.

After housing is installed, turn screw (2) counter-clockwise until bottomed, then clockwise ½-turn. This will adjust the rockshaft only enough to allow operation. Refer to paragraph 284 for further adjustments. Install the negative signal stop (12 through 17) in rockshaft housing and be sure eccentric (12) is located on cam of operating link (9).

284. CONTROL VALVE ADJUSTMENT. While making any of the fol-lowing adjustments, no load should be on the hitch. Load selector lever should be in "MAX" position. Measure from bottom of selector rod (Fig. 308) to finished surface of valve housing as shown. Adjust linkage to obtain approximately 1-3/8 inches distance.

To adjust the rockshaft valve clearance, refer to Fig. 309. Disconnect the link from rockshaft control lever arm and clamp Vise Grip to lever arm as shown. Affix a reference point 10 inches from centerline of lever shaft to be used to measure valve opening clearance.

CAUTION: Avoid lift arm and lift link while making this next adjustment, as they may move rapidly.

Start and idle engine, place load selector lever in "ZERO" position; then carefully measure movement of reference mark required to change rockshaft direction. Movement should be 3/16 to 3/8-inch as shown. If distance is greater than 3/8-inch remove adjusting plug (6—Fig. 302) from side of housing and, reaching through plug hole, insert a screwdriver blade in slotted end of adjusting screw (2—Fig. 303). Turn adjusting screw clockwise to decrease free movement.

Raise rockshaft approximately half way and shut off engine. If rockshaft starts to drop, insufficient valve clearance exists or valves are leaking.

To adjust control lever, remove the load control arm extension plug (Fig. 310) and turn adjusting screw clockwise as far as it will go. With engine running slowly and load selector lever set at "ZERO" position, the rockshaft should just fully raise when control lever is moved rearward so the FRONT edge of lever is even with the stop on the console guide. Adjust lower yoke on the valve operating rod to obtain this setting. Be sure rockshaft rotates a full 76 degrees.

To adjust the negative signal stop, move load selector lever to "MAX" position and run engine slowly. Fully lower rockshaft. If necessary, adjust negative stop screw (Fig. 311) until control lever can be pulled to ¾-inch from rear edge of slot with REAR edge of lever without raising rockshaft. Turn negative stop screw counter-clockwise until rockshaft just starts to raise. This

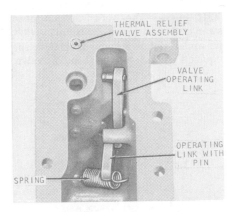

Fig. 305—Assembling linkage in valve housing.

Fig. 306—Valve operating cam and adjusting link installed.

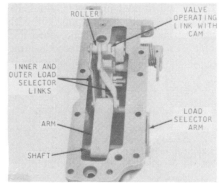

Fig. 307—Fully assembled linkage installed in housing.

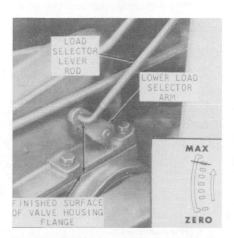

Fig. 308—Distance from rod to finished surface should be approximately 1-3/8 inches with selector lever set at "MAX".

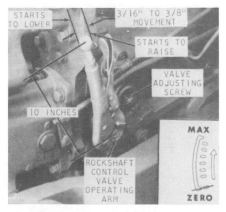

Fig. 309—With load selector lever set at "ZERO", a 10-inch lever should move the distance shown. Valve adjustment is through plug hole in housing.

Fig. 310—Remove plug for access to load control arm adjusting nut.

is the adjustment which prevents the rockshaft from lowering when operating with a reverse load on the load control arm.

285. FRICTION DEVICE ADJUSTMENT. The rockshaft control lever friction adjusting screw is located near the load selector lever (Fig. 312). With the valve operating rod disconnected at the lower operating arm on rockshaft valve housing, the control lever should

Fig. 311—Refer to paragraph 284 for negative signal stop adjustment procedure.

Fig. 312—Refer to paragraph 285 for adjustment of friction device.

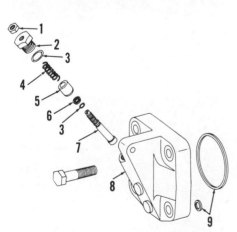

Fig. 313—Exploded view of typical rockshaft cylinder cover and associated parts.

1. Nut
2. Bushing
3. "O" ring
4. Spring
5. Throttle valve
6. Back-up ring
7. Valve shaft
8. Cylinder cover
9. Packing

require a 4-5 lb. pull to make lever move. Loosen jam nut and adjust friction screw to obtain the desired resistance.

286. **ROCKSHAFT HOUSING OVERHAUL.** Rockshaft piston can be removed for inspection or renewal without removal of rockshaft housing. Remove cylinder end cover (9—Fig. 312) and force piston out by pushing down on rockshaft arms with short, jerky motions.

NOTE: Be careful not to damage open end of cylinder with connecting rod or ram (crank) arm. Tighten cylinder cover retaining cap screws to a torque of 170 ft.-lbs. when reinstalling.

To disassemble the rockshaft, remove housing and lower cover (6—Fig. 313). Remove dog point set screw (11—Fig. 315) and right hand lift arm (2) on all except 4630 models, then slide rocker shaft (9) out left side of housing, removing crank arm (8) and servo cam (10) as shaft is withdrawn. The 4630 model is just the opposite, and must be removed out the right side of housing.

When installing rockshaft bushings, use suitable drivers and make sure oil holes are aligned. Add as many spacers (12) as will fit in housing, to eliminate end play. Install servo cam (10) with ramp up as shown and make sure dog point set screw (11) enters locating

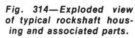

Fig. 314—Exploded view of typical rockshaft housing and associated parts.

1. Oil seal
2. Bushing
3. "O" ring
4. Housing
5. "O" ring
6. Cover
7. Cover
8. "O" ring

Fig. 315—Exploded view of typical rockshaft, piston, lift arms and associated parts. (4630 model shown).

1. Washer
2. Lift arms
3. "O" ring
4. Back-up ring
5. Piston
6. Piston rod
7. Spring pin
8. Crank (ram) arm
9. Rockshaft
10. Servo cam
11. Set screw
12. Spacer washer

hole in rockshaft (9). Splines on crank arm (8) and shaft (9) are indexed for proper alignment during assembly. Tighten lift arm cap screws to 300 ft.-lbs., strike arm with hammer and retighten.

LOAD CONTROL ARM AND SHAFT

All Models

287. **OPERATION.** When the load selector lever is moved to "MAX" position, the operating depth of the three-point hitch is controlled by the draft of the attached implement acting in conjunction with the position of the control lever.

The amount of draft is transmitted by the lower links to the drawbar support (Figs. 319 and 320), then to the control valve by the load control shaft and arm. The spring steel load control shaft is anchored in each side of the transmission housing and the drawbar or draft link support frame affixed to outer ends of shaft. Positive or negative draft causes the center of the load control shaft to deflect a predetermined amount according to the load encountered. The center arc of the flexing shaft moves the straddle mounted lower end of the load control arm while the upper (follower) end transmits the required signal to the control valve.

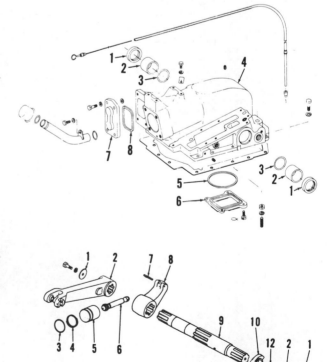

Adjustment is made as outlined in paragraph 289.

288. REMOVE AND REINSTALL. On the 4030 model only, the load control arm can be removed by removing cap screws to support (12—Fig. 316) after first removing rockshaft housing as outlined in paragraph 282. All parts are available individually. On all other models, the differential must be removed, as outlined in paragraph 222.

To remove the load control shaft or draft link support, drain transmission and hydraulic fluid and remove snap rings (Fig. 319 or 320) and retainers (6). Suitably support draft link support, then with a brass drift and working from right side of tractor, bump load control shaft to the left and out. Bushings (7—Fig. 320) in draft link support and (8) in transmission housing can be renewed at this time as can oil seals (10). Renew transmission housing bushings only if necessary. Chill bushings before installation. Inside diameter of bushings is tapered to provide a small bearing area for the flexing shaft (1). Install bushings (8) in transmission housing with small I.D. to inside and bushings (7) in support with small I.D. to outside, away from transmission. On 4030 models only, install oil seal (10) with sharp edge of outer shell to the inside. Lubricate the shaft (1) and install carefully to prevent damage to oil seals. On all other models, refer to Fig. 321. Heat sealing ring in 160° F water to soften. Drive bushing into transmission case with groove to the outside. Collapse sealing ring, and install in groove in bushing with plenty of lubrication after "O" ring is installed. Press sealing ring round so that load

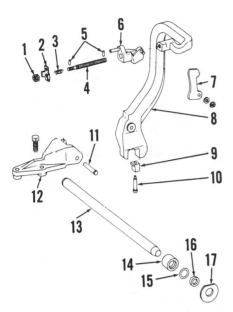

Fig. 316—Typical load control arm and associated parts used on all models of Syncro-Range and Quad-Range (except 4630).

1. Nut	
2. Lock plate	10. Pin
3. Spring	11. Pin
4. Adjusting screw	12. Support
5. Spring pin	13. Load control shaft
6. Extention	14. Bushing
7. Follower	15. "O" ring
8. Arm	16. Oil seal
9. Follower block	17. Washer

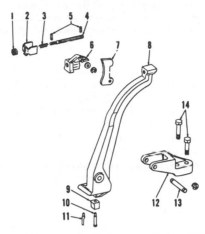

Fig. 318—Load control arm and associated parts used on all 4630 models (Load Control shaft not shown).

1. Nut	8. Arm
2. Lock plate	9. Follower block
3. Spring	10. Pin
4. Adjusting screw	11. Pin
5. Spring pin	12. Support
6. Extension	13. Pin
7. Follower	14. Cap screws

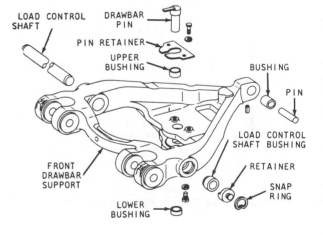

Fig. 319—Exploded view of draft link and front drawbar support used on all models except 4630.

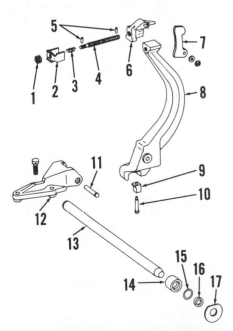

Fig. 317—Typical load control arm and associated parts used on 4230 and 4430 models with Power Shift. See Fig. 316 for legend.

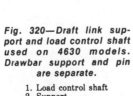

Fig. 320—Draft link support and load control shaft used on 4630 models. Drawbar support and pin are separate.

1. Load control shaft
2. Support
3. Roller shaft
4. Locating screw
5. Roller
6. Retainer
7. Outer bushing
8. Inner bushing
9. "O" ring
10. Sealing ring
11. Washer (selective)

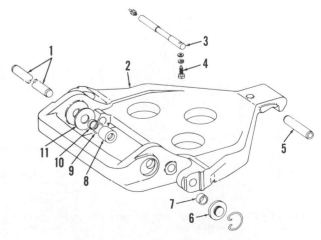

control shaft will go through and install as many selective washers (11—Fig. 320) as necessary to provide minimum clearance between transmission case and support. Carefully install load control shaft with adequate lubrication and install retainers (6) and snap rings. Adjust as outlined in paragraph 289.

289. ADJUSTMENT. To adjust the control mechanism, remove the adjusting plug shown in Fig. 310. Move selector lever to "MAX" and start and idle engine.

Reaching through plug port with a socket and extension, turn slotted nut (1—Figs. 316, 317 or 318) if necessary until rockshaft starts to raise when rear edge of rockshaft control lever aligns with "0" mark at rear of slot on console scale. Check operation of rockshaft and adjust as outlined in paragraph 284.

SELECTIVE (REMOTE) CONTROL VALVES

All Models So Equipped

290. OPERATION. Tractors are optionally equipped with one, two or three selective (remote) control valves for operation of remote cylinders. Mounting positions of valves are shown in Fig. 322.

As with all other units of the hydraulic system, pressure is always present at the valves but no flow exists until the valve is moved. Refer to Fig. 323 for an exploded view of valve mechanism. Each breakaway coupler is equipped with two return valves (20) and two pressure valves so arranged that one of each is opened when control lever is moved off center in either direction. Detent piston (16) is actuated by pressure differential across metering valve (26) and released by pressure equalization when flow stops at end of piston stroke. Flow control valve (25)

maintains an even flow with varying pressure loads.

291. OVERHAUL. Refer to Fig. 323 for an exploded view of the selective control valve. Clamp the unit in a vise and unbolt and remove cap (23) and associated parts carefully as shown in Fig. 324. Identify parts as required for later assembly, then remove valves, springs and guides.

Rotate valve body in vise so that rocker assembly is up as shown in Fig. 325. Rocker arm can be disassembled by driving out spring pin, and removing control arm and shaft. Remove screws holding cam (3 and 7—Fig. 323) and remove rocker (5). Notice how parts are assembled, to aid in reassembly. Inspect all bores, valves and valve seats. Seats are non-renewable, but can

be reconditioned by using NJD 150 Valve Seat Repair Kit (Use exactly as directed). Some tractors were equipped with a spring pin in cover (23) to prevent "0" ring (24) from being pulled into the passage under sudden oil surges.

Assemble by reversing the disassem-

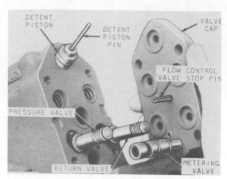

Fig. 324—Clamp housing in a vise to remove rear cap.

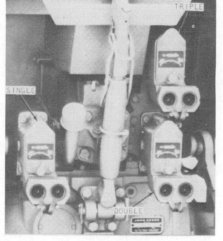

Fig. 322—Rear view of tractor showing location of remote valves.

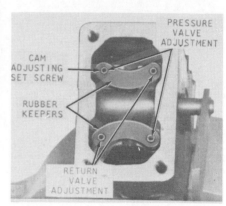

Fig. 325—Front view of housing showing rocker assembly, adjustment screws and associated parts.

Fig. 321—Collapse sealing ring for ease of installation.

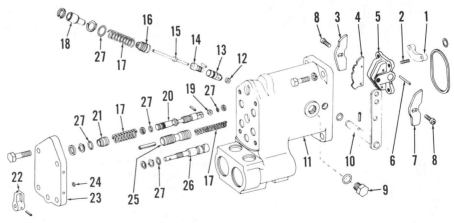

Fig. 323—Exploded view of typical Selective (Remote) Control valve showing component parts.

1. Rubber keeper (2)	8. Cam clamp screws	15. Detent pin	21. Valve guide (4)
2. Adjusting screw (4)	9. Plug (2)	16. Detent piston	22. Lever
3. Regular cam	10. Rocker shaft	17. Spring	23. Cover
4. Detent cam	11. Housing	18. Outer detent guide	24. "0" ring
5. Rocker	12. Detent roller	19. Cam roller (4)	25. Flow control valve
6. Pin (2)	13. Detent follower	20. Poppet valve (4)	26. Metering valve
7. Float cam	14. Inner detent guide		27. "0" ring

bly procedure. If actuating cam was disassembled, refer to Fig. 326 and note that the float cam (7) is shorter than the regular cam (3), and is installed on numbered side of housing as shown. Install detent cam (4) in rocker (5) with 2 pins (6) and install rocker assembly and rocker shaft (10).

Adjusting the valve requires use of special adjusting cover (JDH-15C) and a dial indicator. Remove the two adjusting plugs (9) and loosen the two cam locking screws (8). Back out the four adjusting screws (2) at least two turns.

Install pressure and return valves, detent follower, piston, guides and retaining snap ring. Be sure detent follower roller properly rides on detent cam. Also make sure that operating valve rollers are turned to ride properly on ramps of cams. Back out all adjusting screws on special plate and install the plate with the angled screw pointing at detent pin (15—Fig. 323). Carefully FINGER TIGHTEN the four screws contacting operating valves

until valves are seated; then while holding operating lever in center position FINGER TIGHTEN the detent locking screw until detent roller is seated in neutral detent on detent cam (4—Fig. 326). With operating lever in neutral position, refer to Figs. 325 and 328. Turn in the two diagonally opposite Pressure Valve Cam Adjusting Screws until screws, cams and follower rollers are in contact. Install rubber keeper then back out 1/4-turn as shown. Turn in the two diagonally opposite Return Valve Cam Adjusting Screws until screws, cams and follower rollers are in contact. Install rubber keeper, then back out 1/8-turn. Move the two cams (3 & 7—Fig. 326) into contact with adjusting screws and tighten screws (8) securely.

To double-check the adjustment, install a dial indicator 3 inches from center of shaft on operating arm as shown in Fig. 328. Zero dial indicator while locked in neutral detent, then back out the detent locking screw on adjusting cover. Back out the two

adjusting cover screws which contact operating valves on lever side and measure rocker movement which should be 0.021 inch toward return valve or 0.060 inch toward pressure valve as shown. Valves contacting cam (3—Fig. 326) opposite lever side are being checked. Tighten the two adjusting cover screws on lever side and loosen the other two screws, then check adjustment of valves on lever side. Readjust as necessary for correct rocker movement. This procedure will allow return valves to open before pressure valves, when selective control valve is used.

292. BREAKAWAY COUPLER. Drive a punch into the expansion plugs (17—Fig. 329) and pry out of housing. Remove retainer rings and springs. Operating levers can then be removed. Drive receptacle assembly from housing. Check steel balls, springs, and all parts for wear and replace as necessary. Renew "O" rings and back-up washers. Reassemble in reverse order of disassembly.

POWER WEIGHT TRANSFER VALVE

All Models So Equipped

293. OPERATION. The power weight transfer hitch uses a special coupler, a power weight transfer control valve, pressure gage, a special rockshaft piston cover, a double acting remote cylinder (used as a retracting cylinder only) and a transfer link. The remote cylinder takes the place of the center link on a three point hitch, and becomes a telescoping center link.

When using the power weight transfer hitch, only a drawn implement should be used and operation is in the rockshaft "MAX" position. Excessive load causes the load control arm to di-

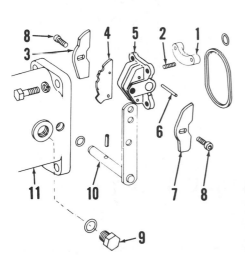

Fig. 326—Exploded view of control rocker. Refer to Fig. 323 for parts identification.

Fig. 328—Use a dial indicator to measure rocker arm movement as shown. Refer to paragraph 291 for procedure.

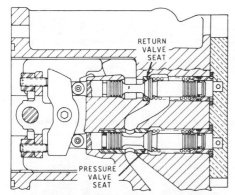

Fig. 327—Valve seats can be reconditioned using special tool NJD 150.

Fig. 329—Exploded view of breakaway coupler and associated parts. All removable parts are duplicated in adjacent bore.

1. Snap ring
2. Snap ring
3. Back-up ring
4. "O" ring
5. Ball
6. Back-up ring
7. Receptacle
8. Ball
9. Spring
10. Plug
11. Snap ring
12. Sleeve
13. Operating lever
14. Cam
15. Lockwasher
16. Washer
17. Expansion plug

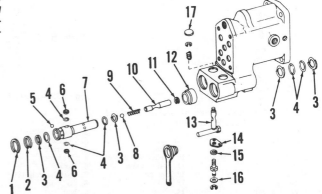

rect pressure oil through the control valve to the remote cylinder. As the cylinder retracts, it tilts the coupler forward at the top, which pulls on the transfer link to the implement being used, causing the weight of the implement to bear down on the draft links. This has the effect of using the rear wheels as a pivot point to unload the front wheels and give added traction to the rear when needed, without having to add ballast to rear wheels. The console mounted pressure gage shows when weight is being transferred.

Refer to Fig. 330 for an exploded view of valve unit and to Fig. 331 for hose routing.

The valve is primarily a switch valve which diverts fluid from the rockshaft cylinder to the control cylinder and rockshaft lever is used for the control lever.

294. **OVERHAUL.** Refer to Fig. 330 for an exploded view of valve unit. Diverting and relief valves can be removed after removing ports plugs (9 & 14). One seat for diverting valve (7) is on upper surface of plug (9); the other seating surface is in bore of body (6). Diverting valve moves upward by pressure of spring (8) and inward flow of oil when knob (1) is backed out for rock-

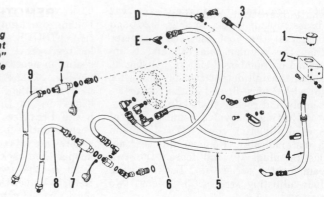

Fig. 331—Hose routing diagram for power weight transfer valve. Fittings "D" & "E" connect to left side of valve.

1. Pressure gage
2. Mounting bracket
3. L.F. port hose
4. Gage hose
5. Right port hose
6. L.R. port hose
7. Coupler
8. Exhaust hose
9. Hose to cylinder

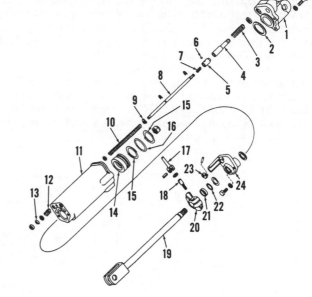

Fig. 332—Exploded view of the high pressure remote cylinder used on all models so equipped.

1. Cap
2. Gasket
3. Spring
4. Stop valve
5. Bleed valve
6. Ball
7. Spring
8. Stop rod
9. Washer
10. Spring
11. Cylinder
12. Spring
13. "V" packing
14. Piston
15. Back-up ring
16. "O" ring
17. Lever
18. Stop screw
19. Piston rod
20. Stop
21. Wiper seal
22. Back-up ring
23. Arm
24. Guide

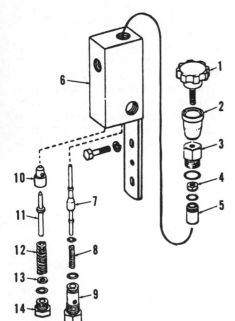

Fig. 330—Exploded view of power weight transfer valve.

1. Valve screw	8. Spring
2. Boot	9. Plug
3. Bushing	10. Relief valve seat
4. Nut	11. Relief valve
5. Shaft	12. Spring
6. Housing	13. Shim
7. Diverting valve	14. Plug

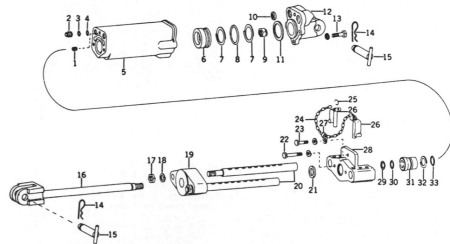

Fig. 333—Exploded view of remote cylinder with mechanical stop. Item 6 may be equipped with an "O" ring.

1. Plug	10. Gasket (2)	18. Lockwasher	26. Stop pin
2. Plug	11. Gasket	19. Piston rod stop	27. Clamp
3. Back-up ring	12. Cap	20. Stop rod (2)	28. Stop rod guide
4. "O" ring	13. Cap screw	21. Oil seal	29. Back-up ring
5. Cylinder	14. Lock pin	22. Cap screw	30. "O" ring
6. Piston	15. Attaching pin	23. Cap screw	31. Piston rod guide
7. Back-up ring (2)	16. Piston rod	24. Chain	32. Back-up ring
8. "O" ring	17. Jam nut (2)	25. Link	33. "O" ring
9. Nut			

shaft operation; and diverter valve seals against upper seat to close off passage to remote cylinder. Turning control knob (1) clockwise mechanically moves diverter valve into contact with seat on plug (9), closing return passage to rockshaft cylinder and opening passage to remote cylinder. Relief valve spring (12) should test 180-220 lbs. when compressed to a height of 1-5/8 inches. Shims (13) may be added if necessary to increase release pressure of relief valve. Renew any parts which are worn, broken or damaged.

REMOTE CYLINDER

All Models

295. Refer to Fig. 332 for exploded view of the double acting, hydraulic stop remote cylinder. To disassemble, remove end cap (1), stop rod spring (3) and valves (4 and 5), using care not to lose ball (6), if cylinder is equipped with override provision. Fully retract the cylinder and remove nut from piston end of piston rod (19). To remove the stop rod and springs, drive the groove pin from stop rod arm (23).

Install new wiper seal (21) in guide (24) with lip to outside, assemble rod end of cylinder, rod packing and piston rings, and have piston fully inserted in cylinder before installing rod nut. Tighten nut securely. Be sure the piston rod stop (20) is located so that stop lever (17) is opposite the stop rod arm (23). Tighten cap screws securing piston rod guide (24) to a torque of 35 ft.-lbs. and cap screws retaining piston cap (1) to 85 ft.-lbs.

Fig. 333 shows parts identification on cylinders with mechanical stop.

NOTES

NOTES

NOTES

NOTES

Technical Information

Technical information is available from John Deere. Some of this information is available in electronic as well as printed form. Order from your John Deere dealer or call **1-800-522-7448**. Please have available the model number, serial number, and name of the product.

Available information includes:

- PARTS CATALOGS list service parts available for your machine with exploded view illustrations to help you identify the correct parts. It is also useful in assembling and disassembling.
- OPERATOR'S MANUALS providing safety, operating, maintenance, and service information. These manuals and safety signs on your machine may also be available in other languages.
- OPERATOR'S VIDEO TAPES showing highlights of safety, operating, maintenance, and service information. These tapes may be available in multiple languages and formats.
- TECHNICAL MANUALS outlining service information for your machine. Included are specifications, illustrated assembly and disassembly procedures, hydraulic oil flow diagrams, and wiring diagrams. Some products have separate manuals for repair and diagnostic information. Some components, such as engines, are available in separate component technical manuals
- FUNDAMENTAL MANUALS detailing basic information regardless of manufacturer:
  - Agricultural Primer series covers technology in farming and ranching, featuring subjects like computers, the Internet, and precision farming.
  - Farm Business Management series examines "real-world" problems and offers practical solutions in the areas of marketing, financing, equipment selection, and compliance.
  - Fundamentals of Services manuals show you how to repair and maintain off-road equipment.
  - Fundamentals of Machine Operation manuals explain machine capacities and adjustments, how to improve machine performance, and how to eliminate unnecessary field operations.

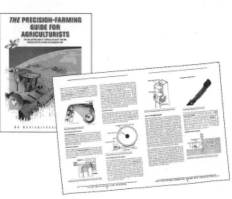

DX,SERVLIT –19–11NOV97–1/1